Quantum Physics for Beginners, Into the Light

The 4 Bizarre Discoveries You Must Know
To Master Quantum Mechanics Fast,
Revealed Step By Step (In Plain English)

John Stoddard

Praise For *Quantum Physics for Beginners, Into the Light*

"This book was so refreshing! It takes the deeply cerebral (and mind boggling) subject of Quantum Physics and explains it in a way that is easy for everyone...If you want an excellent guide through the world of Quantum Physics then get this book...I also will use it as a future reference when questions pop up."

-Perry Duke (US)

"If, as a child, you ever walked through a carnival hall of mirrors and were astonished by how the ordinary suddenly became magical, this is the wonder Stoddard feeds as he examines the twists in Quantum Physics. The book is well organized, both historically and with stepwise spoonfuls of derivations and facts. Either for the beginner as an easily digestible introduction or for the dedicated student as an extra page of practical background and historical trivia, Stoddard is a worthwhile investment."

-Larry Seward (US)

"Everyone interested in Quantum Mechanics should read this! I think the public at large should read this! They will get a greater understanding of the world around them! I definitely couldn't put my tablet down until I finished the book! Absolutely fascinating!"

-John Getty (US)

"This book is a real delight. This book has it all, from the history of modern physics and the major players...to the mathematics and the philosophical interpretations. All of the information is consistent with evidence and scientific consensus too. The introduction told me after reading this I would know more about Quantum Physics than 99% of people, and after completing it I definitely feel like I do...Hats off to the author, truly a gem."

-Joe S. (US)

"This book is a great starting point for those interested in modern physics...Mr. Stoddard eases the reader into the subject; he makes the subject relevant and fascinating. I wish I had a book like this when I was in high school. Mr. Stoddard adds personal touches and humor yielding a book that is easy and fun to read. A book like this is important for modern society."

-Jerry Winniczek (US)

"I thoroughly enjoyed *Quantum Physics for Beginners, Into the Light*...The writing style of text flows quite well: it took me a total of 5.5 hours to read with no fatigue at all. It was definitely not a chore to read like some books...I would say that those 5.5 hours were a very worthwhile investment of my time. My gratitude to the author!"

-Robert S. Van Tassell (US)

"The author lays out in plain, readable and understandable language the meaning and the 'what ifs' of Quantum Theory... It's a book that gives you a sense of wonder and leaves you asking why hasn't anyone else written something like this?"

-J. McCreary (US)

"Great Introduction. Grabs the reader's interest immediately.

The extra-large font for the equations make their importance stand out and helps imprint them on the mind.

Your explanations have an amazing clarity that have expanded my already extensive knowledge of the subject. I especially like the brief anecdotes you use. In particular your description of the evolution of the newer theories – string theory, supersymmetry, M theory, and loop quantum theory is the most concise and informative I have ever read.

The 'Strange Facts' anecdotes are great. They break up the monotony of nothing but text.

Full of interesting and informative quotes by famous physicists.

A glossary at the end is helpful in understanding some of the technical terms."

-J. Dee German (US)

"Quantum Physics is a complicated field of science. It takes a talented person to be able to present the essence of this field in an understandable way to newcomers. The author includes complex equations in the text which may scare any newbie away. Luckily, the author explains the equations in a way that does not leave the beginner feeling lost. This book is a beautiful introduction for anyone curious about his fascinating field of science."

-Andrea LaRochelle (US)

"This is a fun and quick read that gives a great overview of Quantum Physics...The book reviews many aspects of Quantum Physics in a way that is interesting, and easy to understand. I liked that the author included equations and explained them. Also, I thoroughly enjoyed the many 'Strange Facts' that were thrown in as well. Overall, great book that was easy to read! Recommended!"

-Anna Bou (US)

Copyright © 2022 John Stoddard. All rights reserved.

The content contained within this book may not be reproduced, duplicated, or transmitted without direct written permission from the author or the publisher.

Under no circumstances will any blame or legal responsibility be held against the publisher, or author, for any damages, reparation, or monetary loss due to the information contained within this book, either directly or indirectly.

Legal Notice:

This book is copyright protected. It is only for personal use. You cannot amend, distribute, sell, use, quote, or paraphrase any part, or the content within this book, without the consent of the author or publisher.

Disclaimer Notice:

Please note the information contained within this document is for educational and entertainment purposes only. All effort has been executed to present accurate, up-to-date, reliable, and complete information. No warranties of any kind are declared or implied. Readers acknowledge that the author is not engaged in the rendering of legal, financial, medical, or professional advice. The content within this book has been derived from various sources. Please consult a licensed professional before attempting any techniques outlined in this book.

By reading this document, the reader agrees that under no circumstances is the author responsible for any losses, direct or indirect, that are incurred as a result of the use of the information contained within this document, including, but not limited to, errors, omissions, or inaccuracies.

CONTENTS

Introduction   11

Part I
DISCOVERY NO. 1

1. Out With a Bang   19
2. Mankind's Oldest Mystery: The Story of Light   23
3. Planck Pulls a Math Trick, Reluctantly Leads a Revolution   30
4. The True Secret of Light Is Finally Revealed   35

Part II
DISCOVERY NO. 2

5. Shocking Electron Experiment Redefines All Matter Forever   45
6. You'll Never Find the Electron (Here's Why)   50
7. The Most Famous Riddle Ever Told (Answer Included Here)   59
8. The Heart of Quantum Physics   65

Part III
DISCOVERY NO. 3

9. Quantum Field Theory and the Theory of Everything   75
10. The Last Missing Piece of the Cosmic Puzzle   80
11. Einstein Proves Newton Wrong, Sheds New Light on Gravity   85
12. So Where's the Missing Gravity Particle?   91
13. I've Got the World on a String: Promising New Theory May Be the Answer to the Theory of Everything   94

14. The Expanding Universe and the Dark Side of
    the Force   100
15. Einstein Was Right Again, Black Holes Exist and
    We Found One   108

Part IV
DISCOVERY NO. 4

16. Hitler Shocks the World, Einstein Responds   117
17. The Miraculous Phenomenon that Breaks the
    Laws of Physics (And Makes the Atom Bomb
    Possible)   121
18. This Invisible Nuclear Killer Is 10 Times
    Deadlier than the Nuclear Explosion Itself   127
19. How to Build an Atom Bomb, Step By Step   130
20. How to Tame an Atom Bomb (And Lead an
    Energy Revolution)   136
21. What Really Happened at Chernobyl   140
22. All Living Things Have This One Thing in
    Common   146
23. We've Got Most of Superman's Superpowers
    Now (Except Super Disco Abilities)   153
24. The Most Precise Clocks in the Universe   169
25. Hackers Fear This One Technology   175
26. The Next Stage of Computer Evolution   180
27. The Multiverse and Final Thoughts   188

Acknowledgements and Reviews   193
Glossary   195
Equations   207
References   215

*To my father,
who taught me to be curious.*

*And to Ashley, Brittany, and Melanie,
whom I love very, very much.*

INTRODUCTION

> *""I think I can safely say that nobody understands Quantum Mechanics."*
>
> — RICHARD FEYNMAN

"60 seconds left!" the professor barked.

"Till the pain is over," I thought to myself as I stared at the last question on my physics final, a time dilation problem. I didn't know it at the time, but this innocent question would haunt me for years to come.

Question 37. Tony Stark and Thor are in another heated argument that threatens to break up the Avengers forever. To cool off,

Thor boards his ship, leaves Earth, and explores the Galaxy for 10 years according to spaceship time, travelling 95% the speed of light, before returning to Earth. When Thor gets back, how much older is Tony Stark in Earth time?

>   *a. 20 years*
>   *b. 20 minutes*
>   *c. 32 years*
>   *d. 2 minutes*

"To God we pray, let the answer be A." I scribbled down my answer just before the time ran out and submitted the exam, hopeful for a passing grade. A few short days later, the teacher's assistant handed out our graded finals and I was delighted to see I had passed with 83% – but my mind was stuck on something. Not only was I wrong about the time dilation question (Tony Stark aged 32 years, not 20 years), but my natural curiosity about the universe was stirred – if time can slow down, can it speed up? Is it even possible to approach the speed of light? What else was there to learn about the nature of the universe?

Deep down, all of us have pondered these kinds of questions at one time or another. The inner child within each of us looks up at the night sky and yearns to understand these things. Over time, we collect a certain amount of knowledge about the Way Things Are and we hush the inner child as the responsibilities of life steal away our attention. After my physics exam, my mind moved back to the ordinary concerns of life, but the question about Tony Stark stayed with me.

INTRODUCTION | 13

For years, I have nurtured this curiosity and have studied *thousands* of questions of physics in my spare time, crawling forward in my journey inch by precious inch – is there a Theory of Everything? How does time work? Is our universe one among many? Do wormholes exist? More to the point, I have explored a single, overarching question – *what is the true nature of our universe at the most fundamental level*? What are the hidden secrets of light and atoms, underneath everything else?

So, I invite you to join me in this journey of understanding through these pages. I ask you to let your inner child awaken again, and to be open to strange and bizarre discoveries about the true nature of the universe in which you have always lived, even the ones that leave you breathless or frustrated because they make absolutely zero sense at first sight. You will discover:

- **How one young physicist accidentally discovered Quantum Physics with a math trick and changed our entire understanding of light and matter.**
- The true secret nature of light (get ready to be extremely confused).
- **How a playful riddle about a dead cat started as a joke but later smashed our entire interpretation of reality forever.**
- Why and how Newton got gravity wrong (and Einstein got it right).
- **How a man you've never heard of took us to the brink of a Theory of Everything (and why we're not there yet).**

- How the atom really works (hint: the atom you learned about in school is wrong).
- **Einstein's top-secret letter to President Roosevelt that saved the Allies in WWII (and ushered in the modern age).**
- How to calculate the age of dinosaur fossils.
- **The secret of why atomic clocks on GPS satellites run slower than Earth clocks (and how time travel might work).**
- How to build an atom bomb, step by step.
- **Proof there's a black hole at the center of the Milky Way.**
- The secrets behind the world's most powerful computer, which can store more bits than there are atoms in 7.5 billion living humans (your laptop only has 64 bits).
- **The ONE thing in our universe that moves faster than light (this phenomenon spooked even Einstein himself).**
- How we calculated the exact age of the Earth (and the universe too).
- **The deadliest, most destructive threat in our universe (and why we're still alive).**
- How Quantum Physics has built our modern world (and will shape our future).
- **The real science behind worm holes, string theory, alternate realities, parallel universes, and the multiverse... And a whole lot more!**

By the time you're done reading this book, you will absolutely know more about Quantum Physics than 99% of the population (and certainly more than your friends or the film writers at Marvel), but you will *also* have a broader appreciation of Classical Physics and Einstein's Special and General Relativity, in order to provide meaningful context and background to the biggest problem facing physicists today – uniting Quantum Theory with Special and General Relativity to form a unified Theory of Everything. This book does include a modest balance of simply explained advanced mathematics (though it is by no means a substitute for an advanced degree). But if you're like me, you still have a burning curiosity to understand the universe around you, told in simple language - that is the chief purpose of the book you now hold. So, with this modest aim in mind, let us begin our adventure where all good stories start – at the beginning.

PART I

DISCOVERY NO. 1

PLANCK SOLVES THE PROBLEM OF LIGHT WITH A MATH TRICK (AND ACCIDENTALLY BIRTHS QUANTUM PHYSICS)

1

OUT WITH A BANG

"In the beginning, God created the Heavens and the Earth."

— GENESIS 1:1, BIBLE (NIV)

"In the beginning the Universe was created. This has made a lot of people very angry and has been widely regarded as a bad move."

— DOUGLAS ADAMS, "THE HITCHHIKER'S GUIDE TO THE GALAXY"

How did we get here?

Space. Energy. Matter. Time itself. Most cosmologists say that all of it violently and spontaneously burst into being in a massive explosion known as the Big Bang roughly 13.5 billion years ago.

Then, on a tiny, spinning, watery rock orbiting a tiny yellow sun on the edge of a spiral arm of a spinning galaxy orbiting blackhole Sagittarius A, Life emerged roughly 3.8 billion years ago.

And several billion years later, following five mass extinction events including the most recent meteor collision that wiped out the dinosaurs 66 million years ago, our *homo sapiens* ancestors in East Africa crawled out of primitive darkness and became self-aware. Anthropologists estimate this happened around the same time that we mastered the use of fire and tools about 70,000 years ago, in the Cognitive Revolution.

No one knows for certain why or how we became intelligent, and exploring consciousness in depth is well beyond the scope of this book. One theory suggests that our big brains (and weak digestive systems) grew alongside our ability to cook with fire, because less energy was required for digestion (and more energy was available to feed our rapidly growing brains). Another theory suggests that our ability to walk upright freed our grasping hands to create and use tools, spurring faster brain growth. Still another theory suggests that our vocal cords' special ability to produce complex sounds and language enabled us to become the world's most dangerous and effective communicators, cooperators, and predators (Harari, 2011).

Strange Facts: *With our newfound powers, our species quickly spread to every corner of the globe, conquering everything in our path. Archeological evidence indicates we sailed to Australia by boat and exterminated the mega fauna there 45,000 years ago, and that we spread to Europe and drove our cousins, the Neanderthals, to extinction roughly 30,000 years ago. Modern Europeans carry 1-4% Neanderthal DNA today, indicating we interbred with them as well. Our species continues to evolve to this day, generation after generation – the blue eyes mutation, for example, is only 10,000 years old.*

And only 12,000 years ago, in countless, separate places in countless, separate tribes, we invented farming and settled on the land. Freed from the never-ending search for food, our ancestors multiplied and spawned civilizations over the next several millennia. Government and culture flourished. Confucius, Buddha, Jesus, Socrates, the Pharaohs, Aristotle, Alexander, Shakespeare, Caesar, Da Vinci, Newton, and Mozart walked the Earth. Religion, Mathematics, Art, Literature, and Science emerged as we grappled to make sense of the world we had conquered. Mankind had clawed its way to the top of the food chain and possessed the curiosity and freedom necessary to turn its gaze to the skies and ask, among other things, *what are the building blocks of the universe and how does it all work?*

And so we finally arrive to the subject of the book you now hold, Quantum Physics, a young branch of Physics which is just over a hundred years old and seeks to answer that question.

In Classical Physics, we are concerned with the nature and behavior of Big Things. Planets. Stars. Planes, trains, and automobiles. This is the physics of Newton, the physics of the visible and tangible world around us. When we put down our telescopes and begin to observe the world of the very small, however, we enter into the realm of Quantum Physics, the world where particles can pop in and out of existence, particles can exist in multiple states at once, and Newton's laws break down completely. Its goal is to define and describe the properties and behaviors of nature's basic building blocks.

Let us now examine that thing which everyone is familiar with, yet almost no one really understands completely – Light.

2

MANKIND'S OLDEST MYSTERY: THE STORY OF LIGHT

"No great discovery was ever made without a bold guess."

— ISAAC NEWTON

FROM THE AGE OF GODS TO THE AGE OF REASON

Long before early humans understood light, they worshipped it. As early as the 25th century BCE, the ancient Egyptians worshipped the sun god, Ra, whom they regarded as the ruler of all creation and the supreme power of the entire universe. The ancient Egyptians believed that even the Pharaoh himself was just the earthly embodiment of the almighty Ra, and according to their myths, when the other creator gods formed the Earth and Ra explored it for the first time, he was so moved by its beauty he cried deeply - human beings emerged from his tears, of course.

Fig. 1: Ra, the Ancient Egyptian Sun God. Often depicted as a man with the head of a falcon.

In the mid to early centuries BCE, the Ancient Greeks worshipped Apollo, the god of the sun and light, who carried the sun across the sky in his golden chariot each day - his other duties included protecting seafarers and sharing prophecies and instructions from his father Zeus with humans on Earth. He and his brother Poseidon tried to overthrow their father once but failed, and Zeus temporarily stripped Apollo's immortal power as punishment.

Thankfully, Greek science blossomed next to Greek mythology, and the ancient Greeks were the first to develop serious theories about light. While many of them disagreed on the exact nature of light, almost all of them agreed that light is a series of rays which travel in straight lines forever until they bounce off something, and that vision is produced when these rays enter our eyes.

In the 5th century BCE, Plato, the famous Athenian philosopher, wrote that vision requires three "streams of light": one from the object being observed, one from the eyes, and one from the illuminating source. Pythagoras, another famous Greek mathematician and philosopher, believed that light emerges from the eye and provides the experience of sight when it strikes an object. The Greek philosopher Epicurus disagreed with Pythagoras' views on light and put forward his own theory, claiming that objects (not eyes) emit light beams, which then go to the eye and create vision.

Many Arab scholars started working out the process of vision to understand Epicurus' theory. One such Muslim physicist, Hasan Ibn-Al-Haytham, identified the optical components of the human eye and claimed that light isn't emitted from eyes or objects – it is emitted from luminous objects, like the Sun. This light travels from luminous objects and enters our eyes, which then produce vision (Harris & Freudenrich, 2000).

LIGHT IS A WAVE

By the Age of Enlightenment in the 17th century, several renowned European scientists started to think differently about light. In 1690 Christiaan Huygens, a Dutch mathematician and astronomer, proposed an equation describing light as a wave in his *Treatise on Light*, saying that the universe is filled with an invisible material called ether and that light is formed when a luminous body in the ether creates a sequence of waves or vibrations. These waves then go onward until they come into contact with our eyes to produce vision. Even though his equation was flawed, it marked an important step in our understanding (Huygens, 1690).

NEWTON SAYS IT'S NOT

Huygen's wave theory of light was one of the most persuasive wave theories presented in that era. But it was not universally accepted – Sir Isaac Newton himself believed the idea was ridiculous. In 1704, after experimenting and observing the way light reflects and refracts, he concluded that only particles, not waves, could behave in such a way. He called these particles "corpuscles."

Strange Facts: *Newton, though supremely gifted, was a fiercely jealous and paranoid man, especially with his intellectual rivals. Newton scathingly criticized one of these rivals, German mathematician and philosopher Gottfried Leibniz, for years over the question of who had invented Calculus first. Newton accused*

Leibniz of plagiarism and believed he deserved the credit because, although Leibniz published his work first, much of the work was shockingly similar to Newton's own unpublished work. Leibniz claimed to have no knowledge of Newton's unpublished work – this claim was later exposed as a lie.

As an attempt to refute Christiaan Huygen's theory that light was made up of waves, Newton proposed that light is a series of particles, too small and fast to be detected by human eyesight, that moves in straight lines and reflects off objects. His ideas made intuitive sense, and further experiments showing the way light bends, reflects, and refracts seemed to confirm the particle theory – but the discussion was far from over.

NO OFFENSE, MR. NEWTON, BUT LIGHT IS A WAVE

In 1801, Thomas Young published his paper *On the Theory of Light and Color* to the Royal Society – with a simple experiment involving only a piece of metal and a couple slits, he bravely refuted a hundred years on Newtonian beliefs about the true nature of light.

In his experiment, Young took a very thin sheet of metal and cut two very narrow slits parallel to each other. He then blasted monochromatic light through the two tiny holes, which landed onto a screen behind the metal sheet. He then observed something very strange on the screen – a pattern of alternating light and dark spots. He immediately recognized that an interference pattern like this only made sense if light was, not a series of particles, but a series of waves that could converge with other light waves or cancel them out, like water.

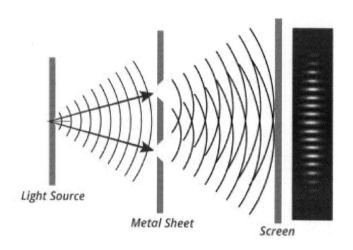

Fig. 2: Double Slit Experiment shows that light is a wave.

Young's research created a new perspective on the light. Scientists began referring to light waves and revised their explanations of reflection and refraction, emphasizing the fact that light waves still obey reflection and refraction laws, like light particles (Parry-Hill, M & W. Davidson, M, 2017).

LIGHT IS AN ELECTROMAGNETIC FIELD?

In 1845, while studying electric and magnetic fields and how they affect each other, Michael Faraday noticed he could induce an electric current by moving a magnet near a wire. The faster he moved the magnet, the greater the voltage produced. He described these electric and magnetic fields and the connections between them in detail and hypothesized that light itself was nothing more than a combination of electric and magnetic fields.

His protegee, Scottish physicist James Clerk Maxwell, carried Faraday's work forward and in 1873, he published his *Treatise on Electricity and Magnetism*, in which he converted Faraday's ideas into their mathematical form. Starting with Faraday's idea that a moving magnet can induce an electric current within a wire, he then showed that the electromotive force (EMF) produced is equal to the line integral of the electric field over a closed loop. This first equation was one of four equations total, Maxwell's equations, a set of related partial differential equations that describe the relationship between electricity and magnetism.

But he didn't stop there. He continued to prove that these electromagnetic signals travel through space as waves, and that they *travel at the speed of light*. He finally concluded that Light is nothing more than a kind of electromagnetic wave (along with radio waves and other radiations). In one bold move, he combined light, electricity, and magnetism into one unifying theory, his theory of electromagnetism (Belendez, 2015).

3

PLANCK PULLS A MATH TRICK, RELUCTANTLY LEADS A REVOLUTION

"Those who are not shocked when they first come across quantum theory cannot possibly have understood it."

— NEIL BOHR, *"ESSAYS ON ATOMIC PHYSICS AND HUMAN KNOWLEDGE"*

Many physicists in the 1890s believed that there was nothing more to say about light and how it operated because Maxwell's theoretical theory of electromagnetic radiation, including its description of light waves, was so elegant and complete. Unfortunately, the description of light as a wave and a wave alone left several questions unanswered – namely, the question of blackbodies.

A blackbody is a special object that absorbs all radiation it receives without reflecting any radiation whatsoever and emits radiation accordingly. A perfect blackbody is nearly impossible to find in nature, but they can be constructed in controlled environments. When a blackbody is heated to a specific temperature, it emits energy with a spectrum that is unique to that temperature, and the intensity of the radiation it emits can be measured very precisely.

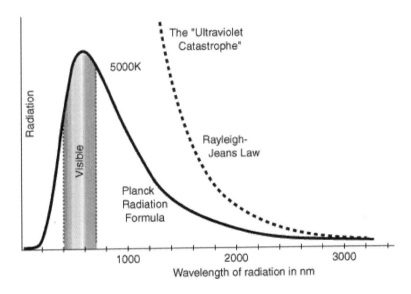

Fig. 3: Black Body Radiation Spectrum across different wavelengths.

Herein lay the problem. The Raleigh-Jeans Law, which viewed light as a wave, predicted that at any given temperature, the radiation a blackbody emitted would be directly proportional to its frequency. As frequency increased (and wavelength decreased), the Raleigh-Jeans law predicted that the blackbody would emit a greater and greater amount of energy, approaching infinite energy. Unfortunately, not only did this idea contradict the **conservation of energy principle** – it didn't match experimental results. The results, shown by Fig. 3 above, showed that as the frequency of light shown on a blackbody increased, the radiation emitted by the blackbody increased as well, but only to a certain point. At extremely high frequencies (or extremely low wavelengths), the blackbody emitted extremely low intensity radiation. Also, at high temperatures of several hundred degrees, the blackbody emitted mostly infrared light, but at extremely high temperatures (such as 5000K as shown in Fig. 3), the blackbody emitted mostly visible light. No one in the scientific community could explain these radiation curves adequately – why weren't they obeying the Rayleigh-Jeans Law? The problem, known as the **Ultraviolet Catastrophe**, remained a mystery. Clearly, there was a larger problem with the way the scientific community was thinking about light (Sherrill, 2006).

PLANCK'S CONSTANT

Max Planck solved the problem. In the early 1900s, the German physicist studied the blackbody radiation curves in depth and worked tirelessly to develop a formula that could describe them accurately, which he finally did. In order to do so, he was forced to let go of his conception of electromagnetic radiation (and visible light) as merely waves. Instead, he conceived of light as broken into tiny packets of energy he called **quanta**, and in the case of blackbodies this energy is released in *discrete packets*, in integral multiples of *hv*:

$$E = nhv$$

Eq. 1: Planck's Equation

In this equation,

E = Energy of the electromagnetic wave
n = an integer or a quantum number
h = Planck's Constant
v = frequency of the light emitted

According to this equation, the energy of a photon is proportional to its frequency by Planck's constant. By precisely measuring the energy released by blackbodies at separate temperatures, Planck calculated his constant to be exactly $6.62607015 \times 10^{-34}$ m² kg / s. He then used this simple formula to derive a complex function showing the radiance of a blackbody as function of its frequency and temperature:

$$B(v, T) = \frac{2hv^3}{c^2} \frac{1}{\exp(\frac{hv}{k_B T}) - 1}$$

Eq. 2: Planck's Radiation Law

In this equation,

B = spectral radiance of a blackbody
v = frequency
k_B = Boltzmann Constant, 1.380649×10^{-23} m² kg/s²k¹
h = Planck's Constant
c = speed of light
T = absolute temperature

And not only did his equation perfectly predict the experimental results - by re-imagining light as made up of quanta, he simultaneously solved the Ultraviolet Catastrophe and birthed an entire new discipline of science, Quantum Physics (Swarthout, 2014).

Strange Facts: *Even though Max Planck received a Nobel Prize for his work with blackbodies in 1919 and is credited as the Father of Quantum Mechanics, he was reluctant at first to accept the idea that light is not a wave, but is quantized. He later commented that he originally employed his idea of quanta as a "mathematical trick" to make his curves match the experimental data. Incredibly, he privately doubted his discovery and fully expected that future experiments would prove him wrong. All future experiments, conducted with ever more precise instruments, confirmed his equations (and the theory underlying the equations) to be accurate. He remains the reluctant Father of Quantum Mechanics.*

4

THE TRUE SECRET OF LIGHT IS FINALLY REVEALED

"The Universe is under no obligation to make sense to you."

— NEIL DEGRASSE TYSON

So Planck showed that electromagnetic radiation is not continuous but is actually broken up into indivisible bits he called quanta, but his ideas were still highly controversial. After all, the double slit experiment showed the world that light behaved like a wave; after Planck's experiments, another young physicist named Albert Einstein would carry Planck's work forward and use another simple experiment involving the **photoelectric effect** to show that light was not just a wave – it was a particle too.

Two decades earlier in 1887, Heinrich Hertz, a German physicist, discovered that when a beam of ultraviolet light strikes upon a metal plate, it causes the plate to spark. Since then, scientists dubbed the phenomenon the photoelectric effect, which occurs when a substance emits electrons when a light beam of any kind strikes its surface at a specific frequency.

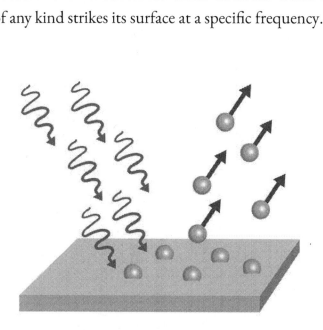

Fig. 4: Photoelectric Effect.

But once again, the scientific community was baffled. According to the wave theory of light, when a beam of light collides with an electron, the force exerted knocks the electrons out of the metal. Therefore, increasing the intensity of the light shining on the metal plate should discharge more electrons at a faster and faster rate with greater and greater kinetic energy. But that's not what scientists observed at all – instead, observers found that increasing the intensity of the light merely increased the number of electrons that were emitted but had no effect on the kinetic energy of the freed electrons. Shockingly, the kinetic energy of the electrons responded to the frequency of the light instead, not the intensity. As the frequency of the light increased, so did the kinetic energy of the emitted electrons. Why did low-frequency, high-intensity light cause no electron emissions, but higher-frequency, low-intensity light did? Once again, the wave theory of light failed to explain this confusing result (Photoelectric Effect Experiment, 2021).

In 1905, a young physicist named Albert Einstein solved this problem by asking, *What if light isn't a wave at all, but a particle instead?* He imagined that light was made up of trillions of tiny particles called **photons.** As these photons collided with the metal plate, the atoms absorbed them and their energy was used to rip the electrons away from the metal – he named the minimum amount of energy required to free the electron the work function. Any remaining energy was transformed into the kinetic energy of the freed electron so that the energy of the photon could be described as the sum of the work function and the kinetic energy of the freed electron.

$$E = KE_{max} + W$$

Eq. 3: Einstein's Photon Energy

In this equation,

E = photon energy
KE_{max} = maximum kinetic energy of the freed electron
W = work function

But his brilliance didn't stop there. Next he used Planck's energy equation to relate the kinetic energy of the freed electrons to the frequency of the light shining on the metal plate:

$$E = hv = KE_{max} + W$$

Eq. 4: Photoelectric Effect

In this equation,

E = Energy of the electromagnetic wave
h = Planck's Constant
v = frequency of the light emitted
KE_{max} = maximum kinetic energy of the freed electron
W = work function

Suddenly the photoelectric effect made complete sense. At low frequencies, the light shining on the metal plate has lower energy and if the frequency is too low, there is not enough energy to free any electrons. As frequency increases, the photon energy increases as well and more and more electrons break free with greater and greater kinetic energy. Einstein's peers, for the first time, conceded that light exhibited characteristics of both waves and particles (Howell, 2017).

Strange Facts: *Einstein later received a Nobel Prize for his work on the photoelectric effect in 1921, but never for his relativity and gravitational theories because the committee was waiting on sufficient proof. Einstein, who had divorced his wife in 1919, shielded his Nobel Prize money from her by placing it into trust for his sons – she could draw interest but couldn't access the capital.*

ARTHUR COMPTON'S SCATTERED X RAYS

Although Einstein made incredible progress in understanding light and even got a Nobel Prize for his work on the photoelectric effect in 1921, many scholars were still convinced that light was a wave and nothing more. In 1923, American physicist Arthur Holly Compton forced the world to acknowledge the particle theory of light all over again, and he did it with nothing more than some X rays, an X ray detector, and a piece of graphite.

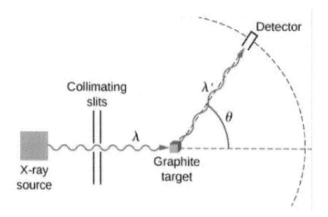

Fig. 5: Compton's experiment shows that light is made of particles.

Compton studied the intensity and wavelength of the incoming X rays in his experiment and measured the frequency and wavelength of the scattered X rays as they bounced off a graphite target. He then adjusted the angle settings of the graphite target and recorded the resulting wavelengths of the scattered X rays at different angles to observe if there would be any changes.

According to the prevailing wave theory of light of the time, he never should have observed any difference - the wavelengths of the scattered X rays should be exactly the same as those of the incoming X rays. But that's not what Compton observed! Instead, he quickly saw that the wavelengths of the scattered X rays were notably longer than the incoming X rays – something very strange was happening.

Like Einstein, Compton imagined the X rays as a stream of photons in order to explain his observations. Once these X rays came into contact with the target material, graphite, they collided with the valence electrons of the graphite, or per electron. The photons scattered from the according to the scattering angle but not before losing some of their energy to the graphite's valence electrons. The scattered X rays contained lower energy and therefore, according to Planck's Law, a lower frequency and a longer wavelength (Goel & Hoang, 2014).

Imagining the X rays as photons, he then derived an equation to relate the difference in wavelengths he observed to the scattering angle and the mass of the electrons inside the graphite:

$$\lambda' - \lambda = \frac{h}{m_e c}(1 - \cos\theta)$$

Eq. 5: Compton's Effect

In this equation,

λ = initial wavelength
λ' = wavelength after scattering
h = Planck's Constant
m_e = electron mass
c = speed of light
θ = scattering angle, typically 0º, 60º, and 120º

Compton, like Einstein, would later receive a Nobel Prize for his work in 1927 for demonstrating the particle nature of light, and his work sent shock waves throughout the scientific community just like Einstein did in 1905. There was no longer any doubt – light exhibits **particle-wave duality.**

Today, physicists describe light as a collection of one or more photons moving through space as electromagnetic waves, acknowledging that yes, light is a wave (and yes, it is a particle).

In this light, how would we redefine Thomas Young's landmark double slit experiment, which clearly showed that light behaves like a wave? A modern physicist would tell you that light travels through the slits as electromagnetic waves and splits once it emerges from the two slits. Once the two emerging wavefields collide with the screen, however, the wavefields collapse as photons on the screen. Light, and indeed the universe, is even stranger than we can imagine (Singh & Quach, 2020).

PART II

DISCOVERY NO. 2

SCHRÖDINGER ASKS A RIDDLE ABOUT A DEAD CAT IN A BOX (AND ACCIDENTALLY SMASHES OUR INTERPRETATION OF REALITY TO BITS)

5

SHOCKING ELECTRON EXPERIMENT REDEFINES ALL MATTER FOREVER

"The double slit experiment has in it the heart of Quantum Physics. In reality, it contains the only mystery."

— RICHARD FEYNMAN

Following the work of Einstein and Compton, the particle-wave duality of light was firmly established –but questions remained. Did matter itself have particle-wave duality too? Could matter behave like a wave? In 1924, French physicist Louis Victor de Broglie re-examined the famous double slit experiment with a slight change in order to answer the question – instead of sending light through the slits, de Broglie fired electrons through them instead and recorded his results.

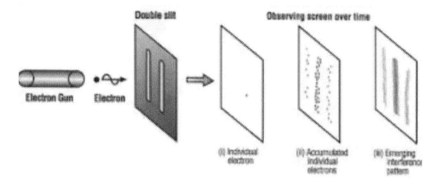

Fig. 6: de Broglie's Double Slit Experiment shows that matter, not just light, has wavelike properties as well.

Incredibly, the results mirrored those of the original light experiment! Large numbers of electrons fired through the two slits invariably produced an interference pattern on the screen, just like waves. Somehow, matter had wavelike properties too (Aharonov, Cohen, Colombo, & Tollksen, 2017).

DE BROGLIE'S MASS-WAVELENGTH EQUATION

De Broglie's next major challenge was to uncover the mathematic principles underpinning his experimental results. He started with Einstein's famous equation which related mass and energy:

$$E = mc^2$$

Eq. 6: Einstein's Mass-Energy Equation

In this equation,

E = energy
m = mass
c = speed of light

Then, he used Planck's equation to show the relationship between frequency and mass:

$$E = hf = mc^2$$

Here, de Broglie made a brilliant observation – Einstein's energy equation dealt primarily with particles moving at the speed of light. de Broglie wasn't interested in relating mass with particles moving at the speed of light. He wanted to show the wavelike properties of *all* matter, which mostly travels much, much slower than the speed of light. With this in mind, he replaced the speed of light with regular velocity :

$$mv^2 = hf$$

Recognizing that the velocity of a particle is the product of its frequency and its wavelength , he rewrote this equation as:

$$mv^2 = \frac{hv}{\lambda}$$

or,

$$\lambda = \frac{hv}{mv^2}$$

Finally, he simplified his equation and irreversibly changed our understanding of matter:

$$\lambda = \frac{h}{mv}$$

Eq. 7: De Broglie's Wavelength Equation

In this equation,

λ = wavelength
h = Planck's constant
m = mass
v = velocity

By slightly adapting the original double slit experiment and applying basic algebra to known and accepted physics theorems, de Broglie was able to show that *all* matter has a measurable wavelength. As his equation shows, the wavelength of matter is impossible to observe with the naked eye, because while particles have extremely low mass (and larger wavelengths), ordinary objects have significantly greater mass (and much, much smaller wavelengths) (Peter, 2008).

Consider, for example, a baseball with a mass of 0.15 kg thrown at a speed of 45 meters per second. Can you calculate the baseball's wavelength off the top of your head? Probably not, but we can both agree it would be too small to notice, even with 20/20 vision.

Incredibly, de Broglie's wavelength theory was confirmed just a few short years later in 1937, when George Paget Thomson won a Nobel Prize for demonstrating the wave-like properties and behavior of the electron. Just as incredibly, George's father, J.J. Thomson, had won the same prize 40 years earlier for proving that electrons existed in the first place.

6

YOU'LL NEVER FIND THE ELECTRON (HERE'S WHY)

"The task is ... not so much to see what no one has yet seen; but to think what nobody has yet thought, about that which everybody sees."

— ERWIN SCHRÖDINGER

So particles have wavelike properties – is it possible to understand them more fully than that? Could we, for example, calculate their exact position over time? In Classical Physics, Isaac Newton's **Second Law of Motion** describes the force of an object in space and time as a function of its mass and acceleration:

$$F = ma$$

Eq. 8: Newton's Second Law

In this equation,

F = force
m = mass
a = acceleration

With this simple equation, it's easy to calculate an object's position over time, so long as we know its initial position and mass (Harris, 2021).

Strange Facts: *Isaac Newton, though regarded in his time as a modern scientist, was a closet Medieval wizard who was secretly obsessed with alchemy, the study of turning base metals into gold. He spent 30 years trying to create gold in his office – what other contributions would he have made to science and mathematics if someone had snapped him out of this obsession?*

In the mid-1920s, although quantum physicists finally grasped the particle-wave duality of matter, there was no "Second Law" to describe the location of particles at the quantum level yet. That is, until an eccentric Austrian philosopher genius named Erwin Schrödinger turned his attention to the field and published his **Schrödinger Equation** in 1926, a partial differential equation which describes the relationship between a particle's momentum, energy, and position. His equation is considered one of the most beautiful and insightful equations ever discovered, not only to the world of Quantum Physics, but to all of modern science.

$$\frac{ih(\frac{\partial}{\partial t})\psi(r,\ t)}{2\pi} = \frac{-(h/2\pi)^2}{2m}\nabla^2\psi(r,\ t) + V(r,t)\psi(r,\ t)$$

Eq. 9: Schrödinger's Equation. It describes both the wave-like and particle-like qualities of a particle as it moves through space and time at slow, non-relativistic speeds.

In this equation,

i = imaginary number, $\sqrt{-1}$
h = Planck's Constant
$\Psi(r,\ t)$ = wave function, defined over space and time
m = mass of the particle
∇^2 = Laplacian operator
$V(r,\ t)$ = potential energy of the particle over space and time

In simple terms, his equation unites the wave nature of a particle with its particle nature and describes its behavior, energy, and location over time. Strictly speaking, if we treat a particle not just as a particle but as a de Broglie matter-wave, then it's impossible to describe its exact location (more on this later) – instead, Schrödinger's Equation is a probability function that gives the *probability* of finding a particle at a certain location given its energy, starting position, and wavelike properties. In a series of four world-changing papers published in 1926, Schrödinger used his wave functions to accurately describe the behavior of a hydrogen atom and gave quantum physicists what Newton had given classical physicists with his Second Law – a mathematical way to accurately describe and predict the location of a particle at a given location or rather, the *probability* of finding a particle at a given location (Aziz, 2020).

The Man Behind the Math: *Erwin Schrödinger's genius and personal quirks were famous, even in his own lifetime. In a time when such arrangements were considered grossly indecent, he kept two lovers his entire adult life, both his wife and his mistress, and he fathered children by both. Even though he grew up in a strict Catholic household, he remained an atheist for life. He fled Germany in the 1933 because he opposed Nazism and the persecution of Jews but later apologized for the move. He made incredible contributions to Quantum Physics but intensely disliked and doubted his own discoveries. Here was a deeply gifted but complicated and troubled man.*

THE ATOMIC MODEL YOU LEARNED IN SCHOOL IS WRONG (HERE'S WHY)

Not surprisingly, Schrödinger's probability density equations clashed with accepted theories of the time, most famously with Danish physicist Neil Bohr's model of the atom, published in 1913. According to Bohr's model, electrons are grouped in concentric circular orbits around the nucleus, depending on their energy level. Each electron has a precise location and follows a distinct path around the nucleus.

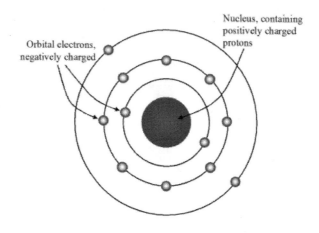

Fig. 7: Bohr's Atomic Model.

Bohr's model, inspired by a dream he had in which Bohr imagined electrons orbiting the nucleus like planets around the sun, was the accepted model of the day (and is still, unfortunately, taught in most classrooms to young children to this day). But Schrödinger applied his wave functions to imagine a *different* vision of the atom, where electrons moved too quickly to calculate their precise location and it was only possible to calculate where they were *likely* to exist. Electrons didn't move in orbits – instead they moved in orbitals, regions of space where his wave functions indicated they were 90% *likely* to be found.

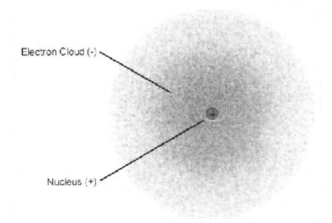

Fig. 8: Schrödinger's Atomic Model.

Schrödinger's work not only defied our theories of the atom – it puzzled and shocked the world because it showed there is built-in *uncertainty* at the quantum level that cannot be fixed with cleverer instruments of observation. We know that electrons exist, but we don't know exactly where they are. An electron could be here or it might be there. It's probably here, and it's probably not there. But we don't know for sure. Over the next several years, Schrödinger's ideas would change physicists' views on the nature of reality itself (David, Harvey, Sweeney, & Zielinski, 2016).

ORBITALS AND SPINS

Particles can exist in more than one place simultaneously, a particle's location and momentum cannot be known at the same time, and a particle can operate as both a particle and a wave (Fore, 2020).

In 1925, Austrian physicist Wolfgang Pauli expanded on Schrödinger's orbitals and proposed that electrons stack up in the orbitals 2 at a time and that the electrons in each orbital have opposite spin:

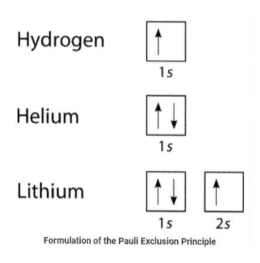

Formulation of the Pauli Exclusion Principle

Fig. 9: Electrons occupying surrounding orbitals in H, He, and Li Atoms.

As the figure shows, the electrons in a hydrogen, helium, or lithium atom, for example, fill up orbitals 1s or 2s in the s shell, and the electrons in these orbitals have opposite spin. According to the Pauli Exclusion Principle, if one of these states is inhabited by an electron with spin one-half, the other can only be occupied by an electron with opposite spin or spin negative one-half (Das, 2022).

Strange Facts: *We've long known that atoms are mostly empty space and that the nucleus accounts for only a very, very tiny piece of an atom's total volume. So if atoms are mostly empty space, why can't we run through brick walls? Why can't our atoms glide through the empty space within the atoms in a brick wall? The answer is the Pauli Exclusion Principle. As electrons stack up in the orbitals surrounding the nucleus, the seats become taken at the low energy orbitals close to the nucleus and the electrons are forced to settle in higher energy orbitals further from the nucleus. This limits how tightly atoms can be packed together and explains why solids are solid.*

Pauli continued to say that each and every electron can uniquely be described by its four quantum numbers, which could be solved with Schrödinger's equations:

1. n is the principal quantum number which describes the position of the electron in the innermost shell.
2. l is the orbital angular momentum quantum number and helps us determine the shape of an orbital.
3. m_l is expressed as the magnetic quantum number, and it reveals the number of orbitals and their orientation.
4. m_s denotes the spin quantum number, and it identifies the direction of the electron spin, up or down.

Formally, the **Pauli Exclusion Principle** claims that no two electrons in an atom can be in the same state or arrangement at the same time. The exclusion principle was later extended to embrace all fermions, including protons and neutrons.

The Pauli Exclusion Principle is essential for governing electron configuration and atom classification in the periodic table, and Pauli won the Nobel Prize in 1945 for his work following a nomination from Albert Einstein himself.

7

THE MOST FAMOUS RIDDLE EVER TOLD (ANSWER INCLUDED HERE)

"[I can't accept quantum mechanics because] I like to think the moon is there even if I am not looking at it."

— ALBERT EINSTEIN

"Nothing is real unless we look at it, and it ceases to be real as soon as we stop looking."

— JOSH GRIBBIN

Schrödinger's wave functions had far-reaching consequences, and physicists later extrapolated his ideas to say that a particle can assume more than one state at a time (the **superposition principle**) and that the particle commits to a single state only once it's observed (the **Copenhagen Interpretation**). In other words, an electron particle, for example, exists in multiple locations and has multiple energy levels *simultaneously*. The act of observation itself causes the set of multiple quantum states to collapse into a single state.

English theoretical physicist Paul Adrien Maurice Dirac described the principle in detail in 1947, saying:

> The general principle of quantum mechanics superposition applies to any dynamical system's states [that are theoretically feasible without mutual interference or contradiction]. It requires us to suppose that there are distinctive interconnections between these states, such that when the system is clearly in one condition, we can consider it to be partially in each of two or more other states.
>
> The initial state must be viewed as the consequence of a form of a superposition of two or more new states in a way that can't be imagined using classical concepts. Any state can be viewed as the consequence of a superposition of two or more other states in an endless number of different ways. Any two or more states, on the other hand, can be superposed to create a new state (Friedman, Patel, Chen, 2000).

THE MOST FAMOUS RIDDLE EVER TOLD (ANSWER INCLUDE... | 61

He further elaborates quantum superposition through an example, cited below as follows:

> Consider the situation where two states, A and B, are superimposed. There exists an observation that, when made on the system in state A, the superposition process is certain to lead to one specific result, say a, and when made on the system in state B, it is certain to lead to a different result, say b (Friedman, Patel, Chen. 2000).

But what will happen if you observe a system that is in a superposed state? The answer is that the result will be either a or b depending on the relative weights of A and B in the superposition process, according to a probability equation. It will never be the same as both a and b (that is, the answer will be one of the two).

This explanation followed the Copenhagen Interpretation of reality – particles exist in multiple states until they are observed.

DISCUSSIONS OF DEAD CATS

Schrödinger despised the Copenhagen Interpretation – a particle can only have a single location or energy level at a time and observation makes no difference, right? In a series of letters he exchanged with Einstein, both grappled with this disturbing interpretation and expressed their mutual desire to uncover another, more complete equation that could calculate an electron's location and energy without employing probability functions. In 1935, Schrödinger published a fuller critique of the Copenhagen Interpretation with a brilliant thought experiment – the famous "Schrödinger's Cat."

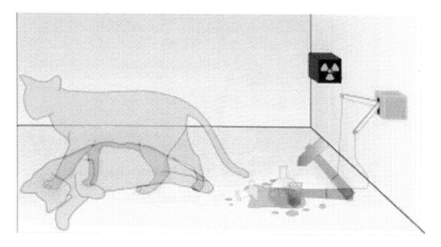

Fig. 10: Schrödinger's Cat Experiment, the most famous thought experiment of all time.

In his thought experiment, he imagined a sealed container containing a cat, a Geiger counter, radioactive material, a hammer, and a deadly poison. The radioactive material had a 50% chance of decaying, causing the Geiger counter to detect it and the hammer to swing and shatter a flask filled with the poison (Baird, 2013).

So, is the cat alive or dead inside the box? Schrödinger reasoned that observing the box had absolutely no effect on the condition of the cat – either it was alive or dead and opening the box would expose its condition, one way or the other. It was impossible for the cat to be both alive and dead at the same time. But several of Schrödinger's contemporaries, most notably Niels Bohr, took a different point of view. Bohr applied the superposition principle to Schrödinger's cat and insisted that yes, the cat was both alive and dead inside the box and that at the exact moment of observation, when an observer opened the box, the superposition of the two states – alive and dead - collapsed so that the cat was observed as either alive or dead. The act of *observing* the cat forced the superposition of states to collapse. Schrödinger's thought experiment, which was intended to show the scientific community how ridiculous this theory was, backfired tremendously, and most physicists uphold the Copenhagen Interpretation of reality to this day. The cat is both alive and dead inside the box, and the cat only adopts one of these states when someone bothers to peak (Merali, 2020).

Strange Facts: *Another competing theory of reality that is not as widely adopted as the Copenhagen Interpretation is the Many-Worlds Interpretation, which holds that the cat is both alive and dead inside the box, but that these live and dead cats exist in separate branches of a multiverse. When the box is opened, the observer may see the cat is alive, but there exists a separate branch of reality in which the observer sees the cat is dead. Every moment of observation is a separate branch point and there are infinite branching realities. We will explore the idea of alternate realities and parallel universes further in the final chapter of the book.*

Fig. 11: Many-Worlds Interpretation, the first modern published scientific theory of multiple realities or parallel universes.

8

THE HEART OF QUANTUM PHYSICS

"Unless something is forbidden, quantum effects and fluctuations will eventually make it possible if we wait long enough. Thus, unless there is a law forbidding it, it will eventually occur. The reason for this is the uncertainty principle."

— MICHIO KAKU, "PARALLEL WORLDS"

Earlier we discussed particle-wave duality, or the idea that particles behave like waves and waves behave like particles. When we described the wavelike behavior of particles, we used de Broglie's equation:

$$\lambda = \frac{h}{mv}$$

Eq. 7: De Broglie's Wavelength Equation

In this equation,

λ = wavelength
h = Planck's Constant
m = mass
v = velocity

Replacing with momentum, we can rewrite de Broglie's equation to describe a particle-wave's *momentum* at any given moment:

$$\rho = \frac{h}{\lambda}$$

We also used Schrödinger's equation to describe a particle's *position* in space, or rather the probability of finding a particle at a particular position in space at a given moment.

In 1927, German physicist Werner Heisenberg submitted his **Uncertainty Principle**, which expands on particle-wave duality and states that although we can measure a particle's position or its momentum at any given moment, we can never measure both simultaneously with certainty. In fact, the more precisely we attempt to measure a particle's position, the more uncertain its momentum becomes, and conversely, the more precisely we attempt to measure a particle's momentum, the more uncertain its position becomes. He realized that this uncertainty has nothing to do with the technological capabilities of our instruments – instead, this uncertainty is built into the fabric of the universe and shatters Newton's conception of a classical universe with precisely measurable and predictable properties to bits (Hilgevoord, 2001).

In *Introduction to Physics and Philosophy* (1962), physicist Paul Davies described the shocking consequences of Heisenberg's work:

> At the heart of the quantum revolution is Heisenberg's uncertainty principle...All physical quantities are subject to unpredictable fluctuations so that their values are not precisely defined...We are free to [measure position x and the momentum p of a quantum particle] to arbitrary precision, but they cannot possess precise values simultaneously...This uncertainty is inherent in nature and not merely the result of technological limitations in measurement (1962).

Heisenberg defined this uncertainty in mathematical terms as:

$$\Delta p \Delta x \approx \frac{h}{4\pi}$$

Eq. 10: Heisenberg Uncertainty Principle

In this equation,

Δ = uncertainty
p = momentum
x = position
h = Planck's Constant

As the equation shows, as certainty in momentum increases, certainty of position decreases and the opposite is true too.

The applications of this principle in the real world are far reaching and explain much of the strange phenomena that has puzzled physicists in the past. For example, logic dictates that in an atom, negatively charged electrons should collapse into the positively charged nucleus – the uncertainty principle, however, shows us that as an electron approaches the nucleus, the error of measuring its position decreases. At the same time, the error of measuring its velocity or momentum increases, meaning that its velocity could be high enough to force the electron to inue its orbit. Another strange example of uncertainty occurs in the process of fusion that enables protons to fuse together to release light and heat from the sun. The temperatures in the sun's core, while high, are not high enough for the protons to overcome their natural repulsion. But the tightly packed protons are relatively still and have an extremely well-defined velocity within the sun, and so their positions are not as well defined. Therefore, there is a small, but non-zero chance that the protons can "tunnel" through the energy barrier separating them and fuse together, releasing massive solar energy. We'll cover this phenomenon in further depth in future chapters on quantum tunneling.

But Heisenberg didn't stop there. He applied his uncertainty principle not just to position and momentum of particles, but to time and energy as well:

$$\Delta E \Delta t \approx \frac{h}{4\pi}$$

This variation of his original uncertainty principle tells us that as time is more constrained, energy is less so and vice versa. This variation explains one of the strangest phenomena in all physics, the random and spontaneous appearance of matter inside perfect vacuums. In a perfect vacuum, there is absolutely zero matter, and yet when modern physicists observe artificial vacuums, they notice over and over again that particles will randomly pop in and out of existence -what explains this? Heisenberg's uncertainty principle applied to time and energy sheds light on the answer – for very, very short, defined periods of time, there is a wildly uncertain energy within a quantum system. This uncertainty of energy is so great that there exists a small, non-zero chance that for tiny intervals of time there can be enough energy in the vacuum to produce particles (Einstein showed us that mass and energy are in fact interchangeable). These particles always appear in pairs, a particle with its anti-matter opposite, and when the particle interacts with its anti-matter opposite, the two **annihilate** each other and pop back out of existence (Melkikh, 2021). We will cover antimatter in more depth in future chapters on Paul Dirac and his incredible prediction of its existence just one year before it was officially discovered in lab experiments. Antimatter's ability to destroy matter itself makes it the deadliest and most destructive threat known in our universe.

Strange Facts: *The problem and danger of annihilation is very real, and modern instruments regularly detect antiprotons in space before they're annihilated. Further, many scientists theorize that the Big Bang should have created matter and antimatter in equal amounts, and that these two classes of matter should have annihilated each other completely, meaning we never should have existed. Their conclusion is that we exist today only because, for every billion matter-antimatter pairs, there was one inexplicable extra matter particle.*

SPOOKY ACTION AT A DISTANCE

Strangely, particles with different quantum states can become linked in a process called **quantum entanglement**, regardless of their uncertain position and momentum. In 1935, Schrödinger and Einstein observed that particles, possessing multiple quantum states simultaneously, may become *entangled* so that their position, momentum, spin, and polarization became correlated *regardless of the distance between them*. In other words, if two particles become entangled and are later separated, we may observe the quantum state of one particle and know with absolute certainty the quantum state of the other, even, theoretically, if the other particle is on the opposite end of the galaxy! For example, when one particle is spinning up following the laws of quantum mechanics, the other particle is always seen spinning down, and vice versa (Cofield, 2017).

The passage of information between the two particles is immediate and travels faster than light, a paradox which Albert Einstein described as "spooky action at a distance." In later chapters, we will discuss exactly how we use this spooky phenomenon in real world technology.

PART III

DISCOVERY NO. 3

UNKNOWN GENIUS SPRINTS TO A THEORY OF EVERYTHING (AND ACCURATELY PREDICTS THE EXISTENCE OF THE #1 DEADLIEST KILLER IN OUR UNIVERSE)

9

QUANTUM FIELD THEORY AND THE THEORY OF EVERYTHING

"Not only is the Universe stranger than we think, it is stranger than we can think."

— WERNER HEISENBERG

By 1920, Einstein had made major leaps in our understanding of the universe with his **general relativity**, which explained a new theory of gravity and spacetime, as well as his **special relativity**, which described the behavior of particles moving at near-lightspeeds. Maxwell had united light, electricity, and magnetism in his theory of electromagnetism, Young and others had discovered the particle-wave duality of light, and Schrödinger had defined wave functions for particles moving at slow, non-relativistic speeds. But Schrödinger's wave functions simply did not apply to particles moving at near-lightspeed, and they didn't describe electron spin, a known phenomenon.

The exciting search was on to uncover a unified "Theory of Everything" – one that would unite the classical physics of big things with tiny things and fast-moving things with slow-moving things. One that would merge all the separate fields of physics into a singular overarching narrative, merge Relativity with Quantum Physics. No one had managed to accomplish this yet, including Einstein himself. There were several missing links.

But in 1931, a young British physicist named Paul Dirac, in a lightning burst of inexplicable inspiration and genius, derived a mathematical equation that broke through the limits of Schrödinger's equation to explain the behavior of an electron moving at near-lightspeeds with spin. *His equation merged all the most important elements of special relativity with Quantum Physics in one stroke* (Gerritsma, Kirchmair, Zahringer, Solano & Roos, 2009).

Dirac's electron equation is written as:

$$(\beta mc^2 + c\sum_{n=1}^{3} \alpha_n p_n)\psi(x, t) = \frac{ih\, \partial \psi(x, t)}{2\pi\, \partial t}$$

Eq. 11: Dirac Equation

In this equation,

Ψ = wave function for the electron with spacetime coordinates x, t
p = momentum
c = speed of light
h = Planck's Constant

Using his equation, Dirac was able to describe the momentum, position, spin, and energy of an electron at near lightspeed and their changes over time, one of the greatest breakthroughs in all of modern physics. Dirac's genius, inspired by Schrödinger himself, was to define his wave function, not as a single electron state, but as the Schrödinger-inspired superposition of four separate electron states: a spin-up electron, a spin-down electron, a spin-up positron, and a spin-down positron. Dirac introduced these new positron particles in order to account for a strange result he could not explain – over and over again, his equation yielded solutions that showed negative energy for the electron. How was this possible? In order to explain his results, Dirac predicted that these negative energies arose from undiscovered particles he named positrons, particles with exactly the same mass as electrons but with positive charge. His equation also predicted that when a positron contacts an electron, the two will annihilate each other, with the combined mass transforming into released energy. Dirac's peers doubted his predictions until physicist Carl Anderson discovered the positron in lab experiments and confirmed Dirac's predictions a year later in 1932 (Webb, 2008).

Strange Facts: *Paul Dirac's electron equation was described by one physicist as "achingly beautiful," and he later won the Nobel Prize for it in 1933. It was rightly regarded as a major step toward a unified Theory of Everything, and it exposed the existence of antimatter even before it was discovered. Known to be eccentric even among physicists, Einstein observed that Dirac constantly balanced on a "dizzying path between genius and madness."*

Dirac's work on electron behavior and antimatter alone would have been enough to immortalize him forever because he forged a link between Quantum Physics and special relativity, but he didn't stop there. In his 1927 paper *The Quantum Theory of the Emission and Absorption of Radiation*, he proposed that photons are not just quantized particles of light – they are also "quanta" of the electromagnetic field. From there, he made an astonishing leap and proposed that *all particles*, not just photons, are nothing more than localized vibrations in a quantum field, and that these fields exist everywhere around us (Pratt, 2021).

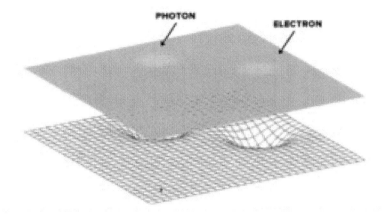

Fig. 12: Localized Vibrations in Quantum Fields. Dirac proposes that all matter, not just light, is quantized.

His **Quantum Field Theory (QFT)** of electromagnetic radiation was another lightning flash of brilliance in the journey toward a Theory of Everything that forced physicists to consider an incredibly new and exciting possibility. *What if every force, not just the electromagnetic force, is composed of quantized fields and that all matter we observe in the universe is nothing more than a vibration in those fields?*

THE LAST MISSING PIECE OF THE COSMIC PUZZLE

"Real gravitons make up what classical physicists would call gravitational waves, which are very weak -and so difficult to detect that they have not yet been observed."

— STEPHEN HAWKING

There are four known forces in the universe: the strong nuclear force, the weak nuclear force, the electromagnetic force, and gravity. In simple terms, the strong nuclear force holds the protons and neutrons inside an atomic nucleus together, the weak nuclear force causes fission and radioactive decay, the electromagnetic force causes oppositely charged particles to attract, and gravity, the weakest force of all, is the attraction between anything with mass or energy (George, 2020).

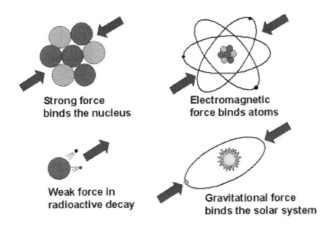

Fig. 13: *Four Fundamental Forces: strong nuclear force, weak nuclear force, electromagnetic force, and gravity.*

Once Dirac published his Quantum Field Theory of the electromagnetic force in 1927, physicists raced over the next several decades to apply a quantum field theory to the rest of the known forces in the universe, and by the mid-1970s, they had built a **Standard Model of Particle Physics**, shown below:

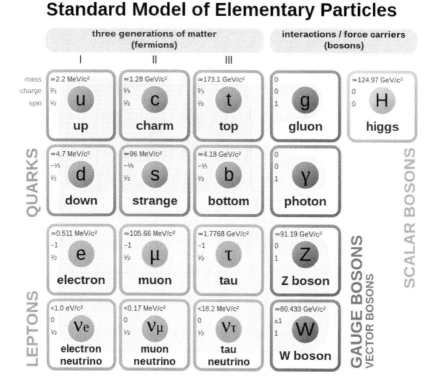

Fig. 14: Standard Model of Particle Physics.

We've already discussed several subatomic particles – electrons, protons, neutrons, and photons. The Standard Model expands on this list and explains that all particles are broken down into two classes – matter particles (fermions) and force-carrier particles (bosons). Fermions are then broken down further into quarks and leptons, which comprise electrons, protons, and neutrons. A proton, for example, is made up of two up quarks and one down quark so that its charge, according to Fig. 14, is

2/3 + 2/3 − 1/3, or +1. A neutron, on the other hand, is made up of two down quarks and an up quark so that its charge is -1/3 − 1/3 + 2/3, or 0. Notice also that an electron has a -1 charge, which means that as long as an atom has the same number of protons and electrons, it will be neutral. Force-carrier particles or bosons carry three of the four known forces in the universe. Gluons carry the strong nuclear force which holds quarks together inside the nucleus and prevents positively charged protons from repelling each other, photons carry the electromagnetic force, and W and Z bosons carry the weak nuclear force (George, 2020). The weak nuclear force drives the beta decay form of radioactivity, which we'll explore in greater depth in later chapters.

Strange Facts: *The Standard Model does more than merely list known particles – it predicts the existence of undiscovered particles. In 1964 physicist Peter Higgs and five other scientists proposed that a "Higgs field" fills all space and that the undiscovered Higgs boson was responsible for enabling mass particles to gain mass at low energies (a necessary phenomenon to explain our existence). This mysterious undiscovered boson, named the "God particle," was finally discovered in 2012 by a team of scientists in Geneva, Switzerland.*

Unfortunately, our current Standard Model, while elegant and beautiful, remains incomplete, and it contains a major missing piece, one that holds us back from our unified Theory of Everything – where is the force-carrier particle for the gravitational force? Where is the graviton? Once physicists discover the graviton, we will indeed have a complete quantum theory for all four of the Fundamental forces in the universe. Alas, the graviton eludes us, and the search for its existence continues even today.

EINSTEIN PROVES NEWTON WRONG, SHEDS NEW LIGHT ON GRAVITY

Let's briefly take a step back to examine gravity, that force which everyone is intimately familiar with yet almost no one truly understands. The great scientist and mathematician Isaac Newton observed an apple falling from a tree and deduced that some force must be acting on the apple to cause it to fall to the Earth. Soon after he gave us his **Law of Universal Gravitation** in his landmark work *Principles* in 1687, which tells us:

$$F = G \frac{m_1 m_2}{r^2}$$

Eq. 12: Law of Universal Gravitation

In this equation,

F = gravitational force between two objects
G = gravitational constant
r = distance between the objects
m_1, m_2 = masses of separate objects

For most instances, Newton's gravitation theory is incredibly accurate, but there were weaknesses discovered later – namely, it failed to explain the behavior of extremely massive or dense objects or the behavior of massive objects at extremely close distances, such as Mercury's orbit around the sun. While Newton's gravitational equations explained the elliptical orbits of faraway planets very well, Mercury's orbit, which varies greatly between 29,000,000 to 43,000,000 miles away from the Sun, remained an anomaly – what was the secret of Mercury's orbit?

In 1915, Einstein attempted to explain these gaps in understanding with his own theory of gravity, **general relativity**. His incredible theory emerged from a very simple idea – *what if gravity was an illusion?* What if the act of "falling" that we observe in the natural world wasn't the consequence of an external force, but something else entirely? In his conception of the universe space and time are connected in four-dimensional spacetime that is all around us. Massive bodies cause curvatures in spacetime that we *experience* as gravity. The curvature of spacetime causes things to gravitate toward massive objects, like a marble rolling toward a bowling ball in a bedsheet. Suddenly, the Mercury problem made complete sense – the Sun's mass was creating a gravity well into which all planets in the solar system were falling, and since Mercury was the closest planet to the Sun, the Sun's gravity well affected it most, causing its orbit to change the most. But curvatures in spacetime don't just create gravity – they also cause light itself to bend as it travels near massive objects (Urone & Hinrichs, 2012).

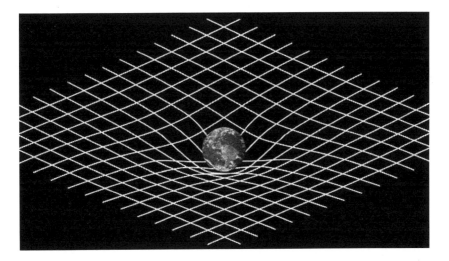

Fig. 15: Earth possesses enough mass to warp the spacetime enveloping it, which we experience as "gravity."

Einstein was so confident he was right that he boldly calculated the exact amount of starlight shift an observer from Earth could expect to see as starlight passed the sun. During a solar eclipse in May 1919, British astronomers Arthur Eddington and Frank Dyson attempted to test Einstein's prediction. They made precise measurements of star locations in the Hyades cluster which were visible during the eclipse from two separate locations, the West African island of Príncipe, and Sobral, in northern Brazil. The team took photographs of the Hyades cluster during the eclipse, and then they compared these photographs to those taken during nighttime, when the Sun is on the other side of the Earth. What they discovered was absolutely incredible – the observed locations of the stars during the eclipse were different from the observed locations of the stars at night, by exactly the amount Einstein's equations predicted. The Sun was so massive that it was curving spacetime and bending distant starlight as it traveled around the Sun to Earth, a process now known as **gravitational lensing**. Additional predictions in Einstein's theory, including the existence of black holes and gravity waves, have been confirmed in the decades since.

And not only does the curvature of spacetime bend light – it also causes *time* to slow down relative to outside observers, and the more mass an object has, the more **time dilation** it will create. Recall the Avengers from the introduction of this book once again, and how we briefly showed that traveling at near lightspeeds causes time to dilate so that time moves more slowly for the Avenger moving at near lightspeed. In this way, Thor could "time travel" into Earth's future by leaving Earth at near lightspeed and returning there later at the end of his voyage. Einstein's general relativity explains that time also moves more slowly near supermassive objects. If one twin remained on Earth while another twin lived on a distant supermassive planet that was much, much larger than Earth, the twin on the supermassive planet would age more slowly than the twin on Earth.

Strange Facts: *Time dilation occurs all around us at all times near all objects with mass, even on extremely minor scales that are difficult to measure. For example, you would experience time more slowly at sea level than you would if you stood at the top of a very high skyscraper. Similarly, if you were orbiting Earth, time would progress faster than if you were on the surface, which explains why clocks on GPS satellites orbiting Earth regularly fall out of sync with clocks on the Earth's surface and must regularly be re-adjusted. Note that large masses and high speeds can alter the rate of time's passage but that it always moves forward – at the time of this publication, backward time travel remains theoretically impossible (sincere apologies to science fiction fans everywhere).*

ON WORMHOLES

Before we continue the story of Quantum Physics and the search for the missing gravity particle, let's briefly examine another one of Einstein's predictions that has fascinated science fiction fans for decades – wormholes. His grand vision of general relativity and curved spacetime, which had been proven in the physical world by Sir Eddington's famous eclipse experiment, led to many logical additional questions. *If spacetime is like a fabric that can bend, what would happen if the fabric folded over itself? Could it be possible to instantly travel vast distances in the universe, simply by moving through these folds in spacetime?* In 1935, Albert Einstein and Nathan Rosen predicted the existence of "Einstein-Rosen bridges," popularly known as **wormholes**, and theorized that objects in space with immense gravity, such as black holes (which will be covered in depth in future chapters), could indeed create temporary wormholes in spacetime. The simplest way to explain a wormhole tunnel is to imagine a piece of paper as the fabric of spacetime. Fold the paper in half and stab a pencil through it so that the pencil pierces both sheets. You have a black hole on one side, through which nothing can escape, and a "white hole" on the other side, through which nothing can enter, with a wormhole joining them. Unfortunately for science fiction fans, Einstein and Rosen predicted two major problems that would render wormholes useless as a means of future space travel. First of all, they would be incredibly tiny and invisible to the naked eye. Secondly, they would also be incredibly unstable, likely collapsing merely moments after formation, because any matter passing through a wormhole would have a gravitational force of attraction that would inevitably pull the walls of the wormhole shut (Redd, 2017).

12

SO WHERE'S THE MISSING GRAVITY PARTICLE?

By the mid-1900s, Planck had set off the quantum revolution with his theory of quantized light, Dirac had carried the torch further with his theory that the electromagnetic force was quantized (and perhaps all forces were quantized too), and Einstein had forever changed our understanding of mass, spacetime, and gravity. But where was the missing link between General Relativity (GR) and Quantum Physics, between the behavior of Big Things and Small Things? Where was the missing force-carrier particle for gravity? If the other forces are quantized, then gravity is likely quantized as well. We've discovered the force-carrier particles for the other three forces, so it follows that "gravitons" exist as well. So, *why haven't we discovered gravitons yet?*

Part of the problem is a practical one – gravity is much, much weaker than the other three forces. For example, in a hydrogen atom, the electromagnetic force between an electron and a proton is 1,039 times larger than the gravitational force that exists between the same two particles.

To illustrate the point further, consider a paperclip latching onto a small magnet. The magnet holds the paperclip tightly in place, despite the fact that the paperclip is also being pulled toward the ground by the gravitational force of an entire *planet*. Gravity, it turns out, is incredibly weak, and in the quantum realm, gravitons are practically impossible to detect. All our modern particle accelerators have failed to find a single graviton thus far (Lincoln, 2014).

Assuming gravitons *do* exist, we have several predictions describing their properties:

1. To have gravity's limitless range, gravitons must be massless.
2. They must be electrically neutral.
3. They can travel at the speed of light.
4. Because gravitons are attractive in nature, their spin must be different from that of protons or electrons.

In addition to being impossible to find, gravitons had other issues as well – according to the mathematics, if ever two gravitons collided, the energy release would be infinite, a sure sign that the mathematics were incomplete or incorrect. Clearly, there was a problem with the way theoretical physicists were thinking about quantum gravity, and perhaps all of particle physics. In the next chapters we'll discuss other exciting theories that have emerged to move around the problem of the missing graviton and unify general relativity with Quantum Physics.

13

I'VE GOT THE WORLD ON A STRING: PROMISING NEW THEORY MAY BE THE ANSWER TO THE THEORY OF EVERYTHING

What if there was a way to describe the collision of these invisible gravitons in a way that wouldn't yield a release of infinite energy? What if the gravitons weren't particles at all, but *strings* instead? In the 1960s, scientists advanced this simple idea, labeled **String Theory**, and proposed that the fabric of the universe is composed of one-dimensional "strings" rather than dot-like particles that can bend and fold as well as branes, sheetlike entities that can travel across spacetime and in higher dimensions. What we observe in the universe as particles are in reality small vibrating strings that twist and turn in intricate ways and vibrate at different frequencies. A photon may be formed by a string of a specific length vibrating at a certain frequency, and a quark may be formed by a separate string vibrating at a different frequency. Undiscovered gravitons themselves are also strings that vibrate at a certain frequency. The theory was incredibly simple and elegant, and it united general relativity and Quantum Physics under a common framework (Tyson, 2020).

But there were several problems with the theory as well.

1. Firstly, it predicted the existence of a hypothetical faster-than-light particle with negative mass, the tachyon, which was capable of breaking the time barrier and destabilizing space.
2. Secondly, it only acknowledged bosons, or force-carrier particles, and completely disregarded fermions, or mass particles.
3. Thirdly, in order to force the underlying mathematics to be correct, string theorists created twenty-two extra dimensions, referred to as hyperspace, in addition to the four we experience in our universe (height, width, depth, and time).

The exciting theory that was initially simple and elegant was in fact incredibly complex, and seemed to make an ever-greater number of bold and unprovable predictions in order to validate its own truthfulness (Pratt, 2021).

By the 1970s, physicists were ready to explore alternative theories, as well as find ways to prove or disprove their theoretical ideas. In order to fix the fermion problem, several string theorists proposed **supersymmetry**, the idea that all bosons have a "superpartner" with exactly the same mass and charge but opposite spin, and all fermions have a "superpartner" with exactly the same mass and charge but opposite spin. For every electron that exists, there is the s-electron, and so on. Superstring theory also proposes ten dimensions as opposed to the twenty-six dimensions included in string theory, and since the first string theory revolution of the 1970s, physicists have created five separate viable superstring theories in addition to supergravity, which has been shown to permit a maximum of eleven dimensions and no higher.

LET'S JUST SMASH THE SUPERSTRING THEORIES TOGETHER

In 1995, physicist Edward Witten made the stunning suggestion that the five separate superstring theories contained several dualities and were in fact just five limited cases of a single overarching theory in eleven dimensions. His **M-Theory** sparked the second superstring revolution and a wave of renewed excitement around the world as scientists confirmed different aspects of his theory. To this day, M-Theory remains our best candidate for a unified Theory of Everything.

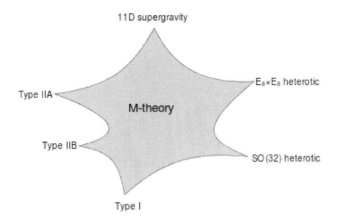

Fig. 16: M-Theory Unites Supergravity and the Five Superstring Theories.

LOOP QUANTUM THEORY

But string theory isn't the only explanation for quantum gravity. In 1986, Abhay Ashketar rewrote the equations of general relativity from start to finish, but instead of imagining particles as points moving in a coordinate system, he imagined them as spinning connections. Soon after, physicists Ted Jacobson and Lee Smolon studied the formal equation of quantum gravity, the Wheeler-DeWitt equation, and realized that when they plugged Ashketar's connection variables into the equations, the equations made sense – their theory of **Loop Quantum Gravity** remains M-Theory's main competitor to this day, and most physicists today fall into one of these two camps.

LQG proposes that the geometry of space is made up of infinite loops woven into a very fine fabric to create a vast network of nodes and links called the spin network. LQG is attractive in its simplicity - it doesn't require several higher dimensions like M-Theory does, and it also does away the problem of singularities, points of matter with infinite density (Zimmerman, Jones, & Robbins, 2016).

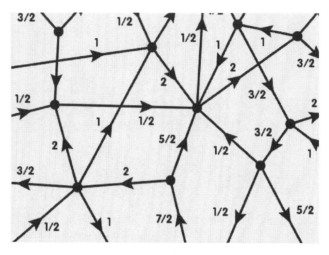

Fig. 17: Loop Quantum Theory Spin Network.

Unfortunately, the theory does contain several flaws – most importantly, it is nearly impossible to prove its accuracy in the real world through experimental observation. Also, the Wheeler-DeWitt equation upon which it is built exists in several varieties and there is no way of knowing for sure which version is the correct one (Zimmerman Jones, A. & Robbins, D. 2016).

As we navigate into the 21st century, we hope to develop these theories further and solve the gravity problem fully. But before we can achieve a proper Theory of Everything, we must first explain three phenomena that complicate our understanding of gravity even further – dark matter, dark energy, and black holes.

14

THE EXPANDING UNIVERSE AND THE DARK SIDE OF THE FORCE

"Your eyes can deceive you; don't trust them."

— OBI-WAN KENOBI, "STAR WARS EPISODE IV: A NEW HOPE"

"Luminous beings we are, not this crude matter."

— YODA, "STAR WARS EPISODE V: THE EMPIRE STRIKES BACK"

In Einstein's time, physicists believed the universe was static, neither growing nor contracting. When Einstein published his Theory of General Relativity in 1915, he realized that the gravitational force of all the matter in the universe should indeed cause the universe to contract. In order to keep the universe stable, he introduced his cosmological constant λ, in order to keep the universe from collapsing from the force of its own gravity.

The physics community regarded the still universe as fact until ten years later, when astronomer Edwin Hubble pointed his telescope to the sky and published his historic paper in 1925, *A Relation Between Distance and Radial Velocity Among Extra-Galactic Nebulae*. First, Hubble studied the behavior of exploding stars, or supernovae, and realized they were rapidly moving away from each other. Next, he plotted the estimated velocities of the supernovae in relation to their distance from Earth and noticed something spectacular – more distant galaxies were receding faster than nearby galaxies. Hubble's Law describes the relationship between a supernova's distance from us and its receding velocity like this:

$$v = H_0 \, d$$

Eq. 11: Hubble's Law

In this equation,

v = receding velocity of a star or galaxy
H_0 = Hubble's constant, 70 km/s/Mpc (where 1 Mpc = 106 parsec = 3.26×10^6 lightyears)
d = estimated distance from Earth

Hubble's Law not only showed that the universe was expanding as well as the exact rate of cosmic expansion – it implied that the universe had been infinitely smaller in the distant past and that its current state was the result of a mysterious Big Bang long ago. By rewriting his Law, Hubble was able to extrapolate different Hubble times, or ages, of the universe for different Hubble constants. Given the present Hubble constant shown in the equation above, Hubble calculated that the universe began its expansion roughly 13.77 billion years ago. *In one leap, Hubble overturned the Static Model of the universe, showed that the universe is expanding, and he calculated its age* (Mann, 2019).

In the decades since then, astronomers have assumed, however, that even though the universe is expanding, the aggregate gravitational pull of the entire universe's stars and galaxies should be slowing its growth. Additionally, it seemed evident that gravity, the force that holds everything together, would eventually put a stop to the expanding universe and it would enter into a future period of contraction (Briggs, 2020).

In the 1990s, two separate teams of astrophysicists looked to distant supernovae to compute this deceleration. Surprisingly, they discovered that not only was the universe expanding but that it was expanding at *a faster and faster rate*. Clearly, there was an unknown force acting against gravity, a mysterious anti-gravity or **dark energy** (Tillman, 2013).

Given the universe's expansion and the force of gravity within the universe, physicists calculated that dark energy must account for about 68% of the cosmos. Truthfully, scientists have never actually observed dark energy or have a clue what it's made of. Nonetheless, scientists have not let their ignorance prevent them from proposing a variety of theories on what this intangible collection of dark energy is, including:

1. A fifth fundamental force in addition to gravity, electromagnetism, and the strong and weak nuclear forces. Some scientists have termed this force or field "quintessence" after Greek philosophers.
2. A fundamental element of space itself
3. A central piece of Einstein's theory of gravity

Until scientists can get closer to validating one of these hypotheses, all we can say for sure about dark energy is that the universe is expanding at a faster and faster rate, and we believe dark energy is responsible for this expansion (Wells, 2021).

ZWICKY'S SPINNING GALAXIES

Eight years after Hubble published his work, Swiss-American astronomer Fritz Zwicky spotted distant galaxies in 1933 and noticed they were spinning around one another far faster than expected, given their observable masses as seen via telescopes. His reasonable conclusion was that the galaxies possessed mysterious, unobserved mass he called **dark matter**, which was causing them to spin at such speeds. "If this is confirmed," he wrote, "we will find that dark matter is substantially more abundant than illuminating matter" (Clavin, 2020).

However, many in the profession remained unconvinced of Zwicky's findings until the 1970s, when astronomers Kent Ford and Vera Rubin conducted thorough investigations of stars in the neighboring Andromeda galaxy's outer regions. These stars were circling the galactic core far too swiftly, almost as if some invisible material was tugging on them and pushing them along – an observation scientists soon detected in galaxies all around the cosmos (Mann, 2020).

Researchers had no clue what this unseen material was formed of. Unfortunately, dark matter is undetectable to the naked eye. Because it emits no light or energy, it cannot be detected by standard sensors and detectors (Tillman, 2022).

Some astronomers hypothesized that dark matter was made up of tiny black holes or other compact objects that emitted too little light to be observed through telescopes. According to

NASA, the results became increasingly odd in the early 2000ss, when a satellite telescope called the Wilkinson Microwave Anisotropy Probe (WMAP) set about observing the microwave radiation left over from the Big Bang. By precisely measuring the microwave fluctuations, the WMAP team was also able to calculate with striking precision the composition of the universe, and they determined that while ordinary matter comprises about 5% of the universe, dark matter comprises 24% of the universe and dark energy comprises 71.4% of the known universe (Briggs, 2020).

WHAT IS DARK MATTER MADE UP OF?

Visible matter, also known as baryonic matter, comprises baryons, which are collective names for subatomic particles such as protons, neutrons, and electrons. Scientists can only anticipate what dark matter is made of. It could be made up of baryons, but it could also be non-baryonic, something different entirely.

Most scientists believe dark matter is made up of non-baryonic stuff. WIMPS (weakly interacting massive particles), the leading candidate, are thought to have ten to a hundred times the mass of a proton. Still, their weak interactions with "regular" matter make them challenging to detect. Neutralinos are another candidate, but they have yet to be discovered.

Sterile neutrinos are another possibility. These are particles that do not exist in ordinary stuff. The sun sends forth a torrent of neutrinos, but they pass through Earth and its inhabitants undetected (Tillman, 2022).

SO WHAT'S THE DIFFERENCE?

Dark energy and dark matter have little in common other than their common adjective, "dark," which doesn't really describe them properly and is only meant to show us that scientists don't fully understand them yet. That said, here are the two main differences between them:

1. Dark matter possesses an attractional force, like gravity, and does not reflect, absorb, or emit light.
2. Meanwhile, dark energy is a repellent force, a kind of anti-gravity, that propels the universe's ever-accelerating expansion (Clavin, 2020).

The significance of dark matter and dark energy is staggering – we live on a tiny spinning rock that is revolving around a very tiny star inside a very small galaxy in a giant universe that is expanding faster than the speed of light. And the trillions of ordinary particles that make up the planets and stars and galaxies inside this universe account for merely 5% of its total mass – the rest of it is hidden from us in the form of dark energy and dark mass. And scattered throughout this canvas are trillions of other invisible and mysterious entities whose gravity is so destructive and so powerful that they threaten to swallow up everything around them and can even warp spacetime itself. We know these entities as black holes. If we can understand the true nature of black holes, we may unlock clues about the nature of gravity and make important progress toward a unified Theory of Everything.

15

EINSTEIN WAS RIGHT AGAIN, BLACK HOLES EXIST AND WE FOUND ONE

"Black holes are where God divided by zero."

— ALBERT EINSTEIN

Incredibly, Albert Einstein himself first predicted the existence of **black holes** in 1916, with his general theory of relativity, decades before our technology developed enough to observe them in the physical world. The term "black hole" was coined many years later by American astronomer John Wheeler in 1967, and today black holes are common knowledge and we find them everywhere in our pop culture.

Black holes started as merely theoretical objects in space until 1971, when the first black hole was spotted by telescope. Just a few years ago, in 2019 the Event Horizon Telescope (EHT) collaboration released the first image ever recorded of a black hole. The EHT spotted the black hole in the center of galaxy M87 while the telescope was examining the **event horizon,** or the area past which nothing can escape from a black hole. Even with this incredibly powerful telescope, black holes are still very difficult to see due to their relatively small sizes and great distances from Earth.

Precisely, black holes themselves cannot be observed. Instead, we are only able to observe the effects and shadows and glow *surrounding* the black holes. These effects include the galaxies and stars which orbit at high speeds around seemingly empty and dark areas of space, the superheated disks of matter that glow due to the immense forces, and the circular shadow produced as the black hole bends light around itself. In 2019, the EHT observed a circular shadow inside galaxy M87, of light bending around a black hole. And in May of 2022, the same EHT finally published an image of Sagittarius A, the supermassive black hole at the center of our own Milky Way. Sagittarius A, located 27,000 lightyears away, is smaller than the size of Mercury's orbit yet contains a mass equivalent to 4 million suns – this discovery directly confirms Einstein's predictions that superdense objects in space, such as black holes, generate enough gravity to bend spacetime and pull in surrounding light, dust, and matter.

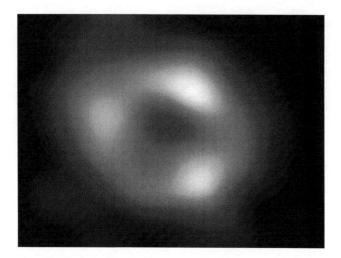

Fig. 18: Black Hole (Sagittarius A) discovered and imaged at the heart of the Milky Way, May 2022.

Strange Facts: *If you fell into a black hole, physicists have long surmised that gravity would stretch you out like spaghetti, though your death would come long before you reached the black hole's center. But a 2012 study published in the journal Nature suggested that quantum effects would cause the event horizon to act much like a wall of fire – you wouldn't be stretched to death. You would burn to death.*

LOOKING AT A BLACK HOLE

So what do black holes look like? Well, a black hole itself cannot be observed and can only be seen as a very dark and featureless area of space. It does not emit or reflect any light. However, there are other aspects of a black hole that can be detected and scientists have a fair idea of the structure. There are two parts to any black hole: the event horizon and the singularity.

The event horizon marks the region surrounding a black hole where gravity becomes so powerful that no light or any other form of radiation can escape. In other words, the escape velocity required to exit the orbit at the event horizon is faster than the speed of light. The event horizon is a sphere that surrounds the singularity which is found in the very center of a black hole. The singularity is more of a mathematical concept rather than something that can be observed. It is the very center of a black hole, a point where all the mass is concentrated. It is infinitely dense and small. The distance between the singularity and the event horizon is called the Schwarzschild radius.

Neither the event horizon nor the singularity is a physical object that can be seen. Instead, we observe the material orbiting the black hole that is pulled in toward the event horizon and ricochets outward, producing powerful jets of glowing material (Briggs, 2022).

THE MIRACLE LINK (AND ITS PESKY PARADOX)

Readers may now appreciate the miraculous importance of black holes – *nothing else in the cosmos unites the physics of Big Things with Small Things, of General Relativity with Quantum Theory, like black holes.* Here is a supermassive object whose gravity is so strong that light can't escape its grasp and entire star systems can't escape its orbit, but also an object which contains a quantum singularity at its center where gravity and density are so high that spacetime itself breaks down. To understand a black hole completely is to understand quantum gravity, and thus uncover a Theory of Everything.

In the 1970s, physicist Stephen Hawking studied black holes closely and made the remarkable discovery that black holes emit Hawking radiation, and that over millions of years a black hole will lose its mass and finally evaporate in a flash of gamma rays. But how is this possible, given that nothing can escape a black hole's event horizon? Hawking showed that near the event horizon, entangled matter-antimatter pairs spontaneously emerged in space (a topic we discussed in earlier chapters on Uncertainty). Ordinarily, these entangled particle pairs would annihilate each other, but near the event horizon, a black hole's gravity would be strong enough to swallow one of the particles in the pair, allowing the other to escape as Hawking radiation.

Hawking's discovery sent shockwaves through the physics community, but it also clashed with one of the most important principles in Quantum Mechanics – that is, the "information" of a particle or a particle system can't be destroyed. So if a particle crosses the event horizon into a black hole, it can't return to us, which is acceptable as long as the information remains intact inside the black hole. But if Hawking's theory is correct, and black holes are ultimately doomed to evaporate through Hawking radiation, then the information of the particles sucked inside them would also evaporate and effectively be lost. This problem, known as the Information Paradox, highlights the fact that we don't yet understand black holes or quantum gravity fully (Tillman, 2021).

FROM DARKNESS TO LIGHT

So far we have covered the most important concepts of quantum theory, including the discovery that light is quantized and that it's both a wave and a particle at the same time. We've also covered Schrödinger's wave-functions, Heisenberg's Uncertainty Principle, Einstein's General Relativity, quantum gravity, and Dirac's Quantum Field Theory. But what do these theories mean for you and me? Why is it important? How has this science changed our world? How will it shape our future? Next, we will answer these questions together and examine the real-world applications and breakthroughs of this incredible science that has built our modern world. We will begin this adventure with the story of the most destructive weapon ever created in the history of mankind – the atom bomb.

PART IV

DISCOVERY NO. 4

EINSTEIN GUESSES HITLER'S SECRET PLANS IN WARNING LETTER TO PRESIDENT ROOSEVELT (AND USHERS IN THE MODERN ERA)

16

HITLER SHOCKS THE WORLD, EINSTEIN RESPONDS

"Now I am become Death, the destroyer of worlds."

— ROBERT OPPENHEIMER. JULY 16, 1945

In August of 1939, Albert Einstein carried the weight of the world on his shoulders. Adolf Hitler, the totalitarian dictator of Nazi Germany, had violated the Munich agreement and invaded Czechoslovakia and was poised to invade Poland as well – peace in Europe and all of western civilization hung in the balance, and although Hitler had struck an agreement with Stalin that he wouldn't invade Poland, Hitler's enemies abroad, including Einstein himself, knew better. In a few short months, World

War II, the deadliest military conflict in human history, would erupt on the world stage.

Just before the Polish invasion, Einstein observed that Hitler blocked the sale of uranium in Czechoslovakia, one of the great sources of uranium ore in the world, and shrewdly guessed that the Nazis were on the path to develop a nuclear weapon. Such a weapon, he knew, would wield destructive power that would eclipse that of any other weapon ever devised and bring its possessor a decisive advantage to any military conflict. His brilliant guess rested on the spectacular discovery of **nuclear fission** of uranium atoms (and its destructive power) in January 1939, mere months before Hitler occupied Czechoslovakia later that year in March.

Einstein immediately penned a secret letter to U.S. President Roosevelt, writing:

> I understand that Germany has actually stopped the sale of uranium from the Czechoslovakian mines which She has taken over…In the course of the last four months it has been made probable…that it may become possible to set up a nuclear chain reaction in a large mass of uranium, by which vast amounts of power and large quantities of new radium-like elements would be generated…This new phenomenon would also lead to the construction of bombs (Groves, 1975).

Einstein then called upon Roosevelt to develop America's own nuclear weapon to oppose the Nazis, and Roosevelt heeded the call, officially placing Robert Oppenheimer as head of the top-secret **Manhattan Project** in 1942. At its height, over 130,000 Americans were working on developing the bomb, spread out across thirty-seven top secret locations in the United States as WWII raged on.

By May of 1945, the Allied Forces, including the United States, Russia, and Britain defeated Nazi Germany in Europe and captured Berlin, but war still continued with Japan. Meanwhile, the scientists of the Manhattan Project were just weeks away from developing a fully operational atom bomb. On July 16, 1945, they detonated the first atom bomb in Alamogordo, New Mexico at a military test facility. The explosion erupted in a brilliant flash of light and radiation, followed by a giant mushroom cloud over the desert floor. Windows of houses more than fifty miles away shattered in the wake of the catastrophic nuclear blast - the nuclear era had begun.

News of the successful detonation of a nuclear weapon reached Truman quickly, but half a world away the Allies still battled the Japanese, who staunchly refused to surrender. Truman's military advisers calculated that a land invasion of Japan would effectively end the war but would come at a heavy price, costing tens or hundreds of thousands of lives of American as well as Japanese soldiers and civilians. They presented the nuclear weapon as an alternative to a costly ground war, Truman accepted, and after Japan refused to respond to Truman's threats of "prompt and utter destruction," Truman authorized the use of the atom bomb on Japan.

On August 6, 1945, an American B-29 bomber dropped the atom bomb on Hiroshima, Japan, devastating the city in an instant and killing 140,000 Japanese civilians. Still the Japanese government refused to surrender. Three days later, on August 9, an American bomber dropped a second nuclear weapon over the city of Nagasaki, Japan, killing over 80,000 Japanese civilians. On August 15, the Soviet Union officially declared war on Japan, and Japan finally issued unconditional surrender, bringing the war to an end.

The atom bomb helped to usher in a lasting era of peace, but not without consequences. Spies in the Manhattan Project, most notably Klaus Fuchs, handed nuclear secrets to the Soviets, and by 1949 the Soviet Union had devised its own nuclear weapon, and a new kind of conflict, the Cold War, spanned the next five decades until the Soviet Union finally collapsed economically in 1991. Nuclear weapons haven't been used in military conflict since 1945 (Kelly, 2005). Still, the *threat* of nuclear war remains the ultimate safeguard of peace between the major powers of the world to this day.

How does the atom bomb work exactly? Before we discuss how to build one, let's briefly cover the physics of the weapon, beginning with radioactivity first and then the science of chain reactions.

17

THE MIRACULOUS PHENOMENON THAT BREAKS THE LAWS OF PHYSICS (AND MAKES THE ATOM BOMB POSSIBLE)

"Unforeseen surprises are the rule in science, not the exception."

— LEONARD SUSSKIND

Radioactive material isn't just built in a lab – it's hidden everywhere in the Earth itself. Most atoms have stable nuclei with balanced forces but there are some cases where an atom may have extra neutrons or protons that carry additional energy, causing the nucleus to become unstable or radioactive. Stable elements are the most common found in nature but unstable forms exist as well – in fact, any nuclei with 84 or more protons will also be radioactive. When an atom is unstable it has a higher energy level and will spontaneously emit particles to move to a lower and more stable energy level until it reaches a stable ground state (Gamow, 1928).

Most radioactive elements cannot decay directly into a stable state, but instead have to undergo a chain of decays until they reach a ground state. All the constituents in this chain are known as the decay series. The uranium decay series sees the unstable uranium-238 undergo fourteen decays before reaching the stable ground state of Lead-206.

Now at this point, precocious readers may be asking a very good question – the strong nuclear force, one of the four known forces, binds the nucleus together and is extremely powerful at close distances. How then is a proton or neutron able to overcome this force to escape the nucleus? The answer to this riddle, according to physicists, lies in a miraculous phenomenon called **quantum tunneling** (Merzbacher, 2002).

Imagine a ball at the base of a hill. The ball is a radiation particle, and the hill is the strong nuclear force that binds an atomic nucleus together. The probability of the ball spontaneously rolling up the hill without any external input is zero, and likewise, the probability that the hill will push the ball back down to the base is one. In other words, a particle has almost no ability to overcome the strong nuclear force. However, according to Quantum Physics, the ball is a particle, but it also has a wave function. As the wave modulates, there is a slim chance that it may have an amplitude large enough to overcome the strong nuclear force, and the ball may roll up and over the hill. This chance is incredibly small – but it is always above zero! Once the ball has successfully passed over this hill, or overcome the strong nuclear force, it will be acted upon by the powerful repulsion of the electromagnetic forces, emitting it at high speeds away from the nucleus.

The miraculous phenomenon of quantum tunneling is what enables particles to overpower the strong nuclear force and escape the nucleus (which means it *also* drives radioactivity of all kinds). Now that we've covered the basic theory underpinning radioactivity, let's briefly explore its three main forms and how they work.

ALPHA DECAY

In alpha decay, the nucleus attempts to find stability by emitting an alpha particle, which is the equivalent of a helium atom and contains two protons and two neutrons. In the chemical equation below, for example, you can see how uranium-238 decays into thorium-234 with the release of an alpha particle. Notice also how mass is conserved at all times in the decay process and the equation remains balanced – both sides of the equation contain exactly 92 protons and 238 protons and neutrons total.

$$^{238}_{92}U \rightarrow {}^{4}_{2}He + {}^{234}_{90}Th$$

BETA DECAY

In beta decay, which is propelled by the weak nuclear force, something very strange happens. One of the neutrons inside the nucleus of an atom will transform into a proton, adding an additional proton to the nucleus and changing the atom into an entirely different element. At the same time, a high-energy beta particle, the equivalent of an electron, is emitted from the nucleus. Below, you can see the chemical equation for how thorium-234 undergoes beta decay to produce protactinium-234 and an extra beta particle. Notice once again how both sides of the equation remain in perfect balance – even though electrons have no protons at all, standard procedure holds that we label them with "-1" protons to maintain the balance. The total number of protons and neutrons on both sides of the equation, 234, remains unchanged.

$$^{234}_{90}\text{Th} \rightarrow {}^{0}_{-1}\text{e} + {}^{234}_{91}\text{Pa}$$

GAMMA DECAY

In gamma decay, a radioactive atom in an excited energy state releases energy in the form of gamma rays in order to reach a lower, more stable energy state. In truth, gamma decay accompanies virtually every nuclear reaction, whether the gamma rays are included in the chemical equation or not. Recall the previous decay of uranium-238 into an alpha particle and thorium-234 – in reality, the decay of uranium-238 also produces 2 gamma rays, which aren't mass particles at all and don't possess any protons or neutrons. Instead, a gamma ray is really a form of high-energy electromagnetic radiation, just like visible light or x-rays. And since we learned from a previous chapter on the Standard Model that everything in the universe is made of mass particles and force-carrier particles and that the photon is the force-carrier particle of the electromagnetic force, you may also hear a gamma ray described as a gamma photon. *Oh yeah...It's all coming together.*

$$^{238}_{92}U \rightarrow {}^{4}_{2}He + {}^{234}_{90}Th + 2{}^{0}_{0}\gamma$$

Strange Facts: *Quantum tunneling not only explains radioactive decay and fission – it also explains fusion, the process that powers our sun. At temperatures above 4 million kelvin and at densities above ten times that of solid lead, hydrogen atoms (single protons) deep in the heart of our sun collide with each other billions of times per second. While most of these protons repel each other, there is a very small chance, about 1-in-10^{28}, that the protons will fuse instead to form helium atoms (two protons and two neutrons), and the resulting energy is released as light and heat. There is a greater chance of winning the lottery three times in a row than there is of witnessing a single successful hydrogen* *fusion, yet there are so many particles within the sun that the miracle happens billions of times per second.*

The miraculous ability of particles to behave as waves and randomly break the strong nuclear force through quantum tunneling is the engine behind radioactive decay, nuclear fission, and by extension the modern atom bomb (Ling, 2016). The radiation released by the atom bomb, as we shall soon see, is incredibly dangerous.

18

THIS INVISIBLE NUCLEAR KILLER IS 10 TIMES DEADLIER THAN THE NUCLEAR EXPLOSION ITSELF

"I do not believe that civilization will be wiped out in a war fought with the atomic bomb. Perhaps two-thirds of the people of the earth will be killed."

— ALBERT EINSTEIN

When an atom bomb detonates, the visible fireball is not the deadliest threat (although I don't recommend standing inside the blast radius). As unstable radioactive elements from the blast decay, they emit harmful radiation that can cause radiation poisoning and death depending on the exposure – even in Hiroshima and Nagasaki, thousands of Japanese died in the nuclear blast, but thousands more died from radiation poisoning in the days and weeks and years following the blast. How does the radiation poisoning work and why is it so dangerous?

When atoms are hit by radiation particles such as alpha particles, beta particles, or gamma rays, they can be stripped of their electrons, causing them to carry a charge, and this ionization process alters the structure of atoms and molecules, causing immense damage to DNA molecules and tissues. These types of ionizing radiation can come from cosmic particles in space, x-rays, and atoms undergoing radioactive decay.

Not all types of radiation can ionize as some do not have enough energy to strip away the electrons but they can still cause the atoms in a molecule to vibrate and move about. These types of radiation are known as non-ionizing radiation and include visible light, microwaves, and radio waves.

Radiation particles also differ in their ability to pass through different kinds and thicknesses of materials, so we can protect ourselves from radiation by using different types of barriers. Radiation particles with high penetration power are considerably more dangerous and difficult to stop. Ionizing power and penetration power have an inverse relationship: particles with a high penetration power will generally have low ionizing power.

For example, alpha particles have the greatest mass out of all the products of radioactive decay. They, therefore, have high ionizing power and can cause significant harm to living tissues, however, their size means that their penetrating power is low and they can be stopped by simple and thin materials like a sheet of paper. The skin offers a significant amount of protection against alpha particles, though some damage may still occur, however, the major risks occur when alpha particles find their way into the body through the lungs or food.

In contrast, beta particles are significantly smaller than alpha particles. They can easily pass through thicker materials - however, a sheet of aluminum is usually sufficient to stop most of them. Despite their higher penetrating power, they have a lower ionizing ability and large quantities would be required to cause comparable tissue damage to the alpha particles (Meredith, 2017).

Gamma rays, unlike alpha and beta particles, carry no mass or charge and are simply a powerful form of electromagnetic radiation. This allows gamma rays to have very high penetration power and only several inches of lead or similarly dense materials can stop them. However, this also means that gamma rays have a low ionizing power compared to other types of radiation particles. This does not make gamma radiation safe though, and the high penetration power means that many more cells and molecules can be exposed to the ionizing radiation, increasing the chances of tissue damage.

The truth is, unstable and radioactive elements are hiding everywhere in plain sight all over the world. One of those elements in particular, uranium, has gained notoriety as the primary ingredient in the deadliest weapon known to mankind – the atom bomb. In the next chapter we'll discuss how to build one.

19

HOW TO BUILD AN ATOM BOMB, STEP BY STEP

"The only use for an atomic bomb is to keep somebody else from using one."

— GEORGE WALD

As usual, physicists stumbled on the science of the atom bomb quite by accident. Before they ever purposely attempted to create explosions with unstable elements, they began with trying to create new ones by firing neutrons at them. In 1934, Enrico Fermi believed that by bombarding uranium with neutrons, the nuclei would become unstable and produce a new element (Amaldi, 2001). Fermi produced two different elements which he named hesperium and ausonium and believed them to have 94 and 93 protons in their nuclei, concluding that he had successfully created new heavy elements through neutron bombardment.

However, Ida Noddack, a German chemist and physicist, was one of the brave few who dared to disagree with this conclusion. Instead, she argued that the neutron bombardment would cause the nucleus to split and form two lighter elements, probably of similar atomic sizes. A few years later in 1939, Lise Meitner and Otto Frisch, recreated these reactions and were able to determine that the products were smaller nuclei. They described this process as **nuclear fission**, in which a heavy element can be split into two lighter elements, usually through neutron bombardment. The nuclei of these elements contain so many protons that it only requires a small amount of energy, such as the collision of a single neutron, to cause a complete breakup of the nucleus. Below is the chemical equation for the fission of uranium-235:

$$^{235}_{92}U + ^{1}_{0}n \rightarrow 3\,^{1}_{0}n + ^{92}_{36}Kr + ^{141}_{56}Ba + ENERGY$$

The secret of the atom bomb is to fire a neutron at a certain quantity of uranium and trigger a chain reaction. When a neutron collides with the nucleus of a uranium atom, two elements are produced, krypton and barium, along with three additional neutrons and massive amounts of energy. Fission reactions release much more energy than natural radioactive decay, and often the products that are produced are unstable.

The additional neutrons that are produced can easily cause a chain reaction to occur as they continue to slam into other atoms and cause trillions of other nuclei to undergo the same fission process (Bonolis, 2001).

At this point, careful readers are wondering, "But atoms are made up of mostly empty space with a very, very tiny nucleus at the center – how are we able to guarantee that a stray neutron will successfully hit another nucleus?" The solution to this problem is to assemble just enough uranium to achieve a high enough density to guarantee successful fission, and to only bring the two halves of this **supercritical mass** together at the exact moment of detonation. To achieve supercriticality, the fuel is either compressed to increase its density using chemical explosives (known as the implosion method) or by shooting sub-critical materials into each other (known as the gun method). This is required because not all of the uranium or Plutonium fuel materials will undergo fission reactions, and some neutrons will be lost to the system. Compare this to the fuel used inside a nuclear reactor, which is only at critical mass. Fuels at critical mass have an equal rate of neutron loss to the rate of neutrons created by fission, and a chain reaction will occur in which one neutron will, on average, be able to initiate another fission reaction. In a supercritical mass, the rate of neutron loss is lower than the rate of neutron creation through fission. So further fission reactions can occur at an exponentially increasing rate. Systems can also exist in a sub-critical state, where the rate of neutron loss is greater than the rate of neutron creation by fission, and so the fission reactions will come to a stop as there are not enough free reactions to cause a chain reaction.

The critical mass of uranium-235 is approximately 104 pounds (47 kg), and for Plutonium-239, the critical mass is approximately 22 pounds (10 kg). The fuels must first be enriched before they can be used in a nuclear reactor or inside an atomic bomb. This is because natural materials come with impurities, and the percentage of uranium-235 inside a sample of uranium will be less than 100%. Impurities can absorb neutrons without undergoing fission reactions. Weapons-grade uranium is generally enriched to about 85% uranium-235, though less efficient fuels can also be used.

Fuels are enriched using lasers to separate the different isotopes. There are three different isotopes of uranium, uranium-235, uranium-234, and uranium-238, which are the most common and abundant. Uranium-235 is the only isotope that can undergo fission, and so the other isotopes must be removed before use in a bomb or reactor.

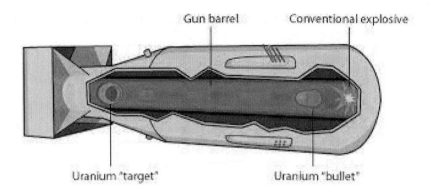

Fig. 19: Design of the Atom Bomb.

The design of the atom bomb is very simple. Two subcritical masses of uranium are at either end of a gun barrel inside the bomb. One uranium mass is shaped like a bullet, while the other uranium mass is a hollow cylinder target that fits around the bullet. Regular explosives shoot the "uranium bullet" down the gun barrel, where it collides with the uranium target, and when the two subcritical masses collide to form supercritical mass, the fission reactions begin occurring in an exponentially growing chain reaction. This sudden chain reaction releases massive amounts of heat and energy, capable of reaching several million degrees and producing shockwaves that can travel several miles. The heat creates a large fireball capable of incinerating towns and cities, which rises upwards and forms the characteristic mushroom cloud. One of the most important design factors is to ensure that all of the fuels can undergo fission before the weapon can destroy itself by becoming too hot and melting.

Fission products, including radioactive material, neutrons, and gamma rays, are also produced when a nuclear bomb explodes. This "fallout" can be carried by winds across hundreds of miles, contaminating environments with materials that can cause radiation sickness and cancer for several years or even hundreds of years after the explosion.

But nuclear fission can do more than decimate cities – it can also power them. In the next chapter we'll cover how that works and why it matters.

20

HOW TO TAME AN ATOM BOMB (AND LEAD AN ENERGY REVOLUTION)

Put simply, nuclear energy is just fission with a carefully placed muzzle on it. The massive amounts of energy produced through nuclear fission reactions are harnessed in nuclear power plants to generate electricity. Critical fission reactors are the most common type. They use neutrons produced by the fission of fuels such as uranium to initiate controlled chain reactions. Other kinds of reactors include subcritical reactors, which don't rely on these chain reactions and instead use radioactive decay or particle accelerators.

Nuclear reactors use radioactive fuel, most often enriched uranium-235, packaged into small pellets stacked into tubes. The fuel is used to produce fission reactions. However, the neutrons produced must be slowed down for a controlled chain reaction where the heat energy can be harvested for electricity production. This is why nuclear reactors need moderators—

materials that can slow down neutrons. The most common type of moderator is water; however, graphite can also be used.

Coolant travels through the reactor core, where it absorbs the heat from the fission reactions, and then travels to an external water source to produce steam which drives turbines that produce electricity. The final part of a nuclear reactor is the containment vessel. This powerful radiation shield protects workers, equipment, and the environment from radiation. Containment vessels are usually made from a combination of steel and concrete. Nuclear reactors can carefully control or stop the chain reactions by partially or completely inserting control rods into the fuel. The control rods strongly absorb neutrons from the fission reactions, preventing further fission reactions from occurring (Waltar, 2003).

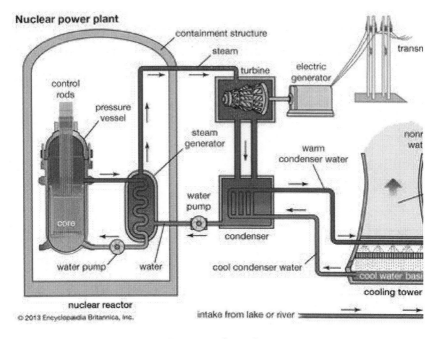

Fig. 20: Diagram of a nuclear reactor.

Two major issues exist with nuclear fission reactors. The first is nuclear waste. This includes the water used as a modulator within the reactor, as well as radioactive fragments produced by the fission reactions. These waste products can remain dangerously radioactive for very long periods and need to be kept away from human contact and contact with the environment as much as possible. The waste may continue to give off heat as the radioactive elements continue to decay.

The second issue is the possibility of a reactor meltdown. Though reactors are designed in a way that makes this extremely rare, it may still happen and has happened before. A reactor may experience a meltdown if the fission reactions inside the fuel rods are not controlled properly, causing extreme temperatures that can melt the rods, and the surrounding infrastructure, allowing the radioactive materials to leak into the environment. Another possible disaster can occur if the control rods are not used properly, and excess neutrons escape. In the next chapter, we'll explore the greatest nuclear catastrophe of all time, the meltdown at Chernobyl.

21

WHAT REALLY HAPPENED AT CHERNOBYL

"Every lie we tell incurs a debt to the truth. Sooner or later, that debt is paid."

— VALERY LEGASOV, *CHERNOBYL (HBO)*

The nuclear disaster that occurred at the **Chernobyl** nuclear power plant near the city of Pripyat in the Soviet Union on April 26, 1986, remains one of the most infamous, catastrophic, and tragic manmade disasters in all of human history. It is still the worst nuclear disaster ever, killing more people and costing more in damages than any other subsequent or prior event. It is the only accident in the history of nuclear power where radiation-related deaths occurred at a commercial facility.

The accident occurred during the Cold War during the midst of a nuclear arms race between the United States (US) and the former USSR. The extreme competitiveness between these two nations fostered a climate of lies and cover ups in all aspects of their nuclear programs, not only between the nations but within the Soviet Union itself. Many Soviet nuclear power plants claimed to have outstanding performance - however, many were notorious for their lack of safety standards, accountability, and technological advancements, especially compared to the US. The USSR also used different types of nuclear reactors than much of the rest of the world, namely the WWER pressurized light-water reactors and the RBMK graphite-moderated light-water reactors.

The Chernobyl power complex had four different RBMK ("Reaktor Bolshoy Moshchnosty Kanalny") nuclear reactors, known as units. Inside the RBMK reactor, the fuel assemblies are all contained in individual pressure tubes designed so that each one can be loaded or unloaded without having to shut down the reactor. This makes RBMK reactors suitable for energy production and Plutonium production. This is in contrast to some of the designs seen in the US, like the LWR reactors, which have one pressure vessel that contains all of the fuel assemblies. The RBMK reactor uses graphite rods as the moderator and water as the coolant, while the LWR reactor uses water as the moderator and as the coolant. If the LWR reactor runs out of coolant for some reason, there will be no moderator material, and the fission reaction will cease.

The use of graphite is one of the many reasons the Chernobyl incident occurred. Inside the RBMK reactor, energy production is controlled by raising or lowering the control rods into the moderators. This increases or slows down the fission rate by absorbing fewer or more neutrons, respectively. One of the problems with this design is the coolant, water, which also has moderating properties. When the coolant temperature rises, it can begin to boil, forming steam bubbles or 'voids.' Steam does not have the same neutron moderating abilities as liquid water, and the rate of fission reactions will increase, creating what is known as a positive void coefficient.

On the day of the Chernobyl accident, workers carried out a safety test to determine how long the turbines would continue to spin and produce power if the steam generation stopped and test the performance of the backup diesel generators. This meant that power production would be gradually reduced under controlled conditions. The test had already been attempted in the previous year, but the reactor shut down too quickly to produce any conclusive results and needed to be repeated.

However, something unexpected occurred. One of the byproducts of the nuclear fission reaction is xenon-135, a material with neutron-absorbing properties that could moderate further fission reactions. Under normal conditions, the xenon would be burned off by the heat of the fission reactions. During the safety test, there was not enough heat being produced to burn off the xenon, and it began to build up, absorbing more and more neutrons and slowing down the rate of fission reactions in a process called reactor poisoning. This put the safety test in jeopardy, as they needed the reactors to slow down at a controlled rate.

At the same time, the coolant pumps had lost power as they were supplied by the turbine, which was shutting down for the safety test. The diesel generators were supposed to help control the pumps, but they did not do so correctly, and coolant began to heat up and form steam voids, creating a dangerous positive void coefficient and an inability to control the rate of fission.

To increase the energy production and make sure the safety test was viable, the workers removed several control rods from the reactor core. This helped burn off some of the excess xenon that had built up, and energy production began to increase. The reactor was volatile, though, with large temperature fluctuations and irregular coolant flow.

The workers knew they had to initiate an emergency shutdown at this point to prevent a meltdown as the core became more unstable. When the workers pressed the emergency shutdown button, all of the control rods raised earlier were reinserted into the core to absorb excess neutrons and stop nuclear fission. *But as the rods re-inserted, they displaced some of the moderating water, causing a dangerous spike in energy production that overheated and damaged the control rods inside the reactor core.* The reactor output jumped to about 30,000 MW during this emergency shutdown - ten times more than its standard output.

The coolant continued to boil away, producing steam that eventually ruptured the pressurized containment channels. This initiated an explosion that damaged the containment vessel, leading to a loss of coolant being pumped through the system, causing an extremely high positive void coefficient that further exacerbated the energy production from the core. A second horrific nuclear explosion completely destroyed the core, ejecting fuel, control rods, and other radioactive material into the atmosphere. Much of this material was carried away as dust and smoke, traveling by wind over large areas of Ukraine, Belarus, and Russia.

The Girl Who Lived: *Janina Scarlet was just three years old when the nuclear disaster at Chernobyl occurred 180 miles from her hometown in Vinnitsin, and she suffered severe seizures and migraines in the months that followed. Two years later, the radiation poisoning came to a head, and several young people in her town began getting cancer and dying. She describes her own suffering: "So many blood vessels had popped that my eyes were completely red and I collapsed." Her family remained in Vinnitsin for several more years till they finally moved to Brooklyn, New York in 1995, where she made a full recovery.*

The fire prevented emergency personnel from getting too close to the reactor site for several hours, which may have been good. However, they and other concerned citizens quickly began arriving at the site to see what had happened, many walking amongst the debris from the explosions. This debris was all highly radioactive, with some pieces resulting in immediate radiation burns. Many of the initial casualties of the event were the firefighters and emergency first responders who were able to extinguish the fire, as well as the plant workers. The next step in the cleanup process involved collecting and removing the debris from the explosion site. More than 200,000 "volunteer liquidators" from the USSR were tasked with this job, and they were exposed to excessively high levels of radiation. The city of Pripyat was evacuated, displacing 45,000 nuclear refugees who were often denied entry to foreign countries over fear of nuclear contamination. About five million people lived in areas that were now considered contaminated, and a 30km radius exclusion zone was set up (World Nuclear Association,2021). The horrific disaster at Chernobyl cost an estimated $235 billion in damages and hastened the collapse of the Soviet Union a few short years later in 1991. To this day, Chernobyl is still considered highly radioactive, and the exclusion zone is still present. Today's nuclear reactors have been modernized in the wake of the Chernobyl disaster, and nuclear energy remains a powerful energy alternative to non-renewable fossil fuels.

But quantum technology extends far beyond bombs and nuclear reactors. Lasik eye surgery. Google Maps. The end of all voter fraud. Miracle computers that take only seconds to solve problems which would take ordinary computers hundreds of thousands of years to solve, if ever. Cybersecurity and modern warfare. Cars that drive themselves. In the next several chapters we'll briefly cover some of the most exciting and world-altering applications of Quantum Physics, which touch nearly every aspect of our lives.

22

ALL LIVING THINGS HAVE THIS ONE THING IN COMMON

"Age is an issue of mind over matter. If you don't mind, it doesn't matter."

— MARK TWAIN

In previous chapters, we've focused solely on the *destructive* powers of quantum science and its ability to end life – on the coin side, quantum science is present in all living things as well.

All living thing things age and die. And because all living things contain carbon, and because Carbon-14 decays at a very specific rate, and because we can compare carbon amounts in fossils to the carbon amounts in modern living things, we can calculate the age of those fossils and uncover the history, piece by piece, of Life on this planet.

The science is simple, really. Let's say I gave you a tray of brownies from the oven that was 8 in long and 8 in wide or 64 sq in. But let's say I warned you that every hour, half the brownies would be eaten, and you needed to grab one quickly. Let's say you fell asleep and woke up and came to the kitchen to see there was only one piece left that was 4in long and 4in wide or 16 sq in – how many hours ago did they come out of the oven? In other words, how "old" are the brownies? The correct answer is that the brownies *are two hours old* (after one hour there was were 32 sq in and after the second hour there were only 16 sq in).

In nature, physicists use the concept of **half-life** to solve this same kind of problem. A 'half-life' is the amount of time it takes for different kinds of radioactive elements to decay by half their mass. Nuclei that have short half-lives are more radioactive and will decay more rapidly compared to nuclei with long half-lives. The rate of decay is directly proportional to the number of remaining nuclei and decreases exponentially over time. Half-lives can range anywhere from 0.16 milliseconds as with polonium-214, to billions of years such as uranium-238 or thorium-232. Half-life is not an exact measure of decay rates but is based on the probability of a radioactive nucleus decaying by 50% within its half-life.

Imagine there are 1 million atoms of a radioactive isotope that has a half-life of ten years. After ten years, there will only be half of the atoms left, about 500 000. After another ten years, this number will half again to about 250 000. As you can see the rate of decay is completely independent of the age of the atoms, and they will continue to half in number every 10 years.

Tritium, a radioactive isotope of hydrogen, decays very fast and has a half-life of only 12.5 years. This means that the probability of tritium decaying within its half-life is 50%. The question is, if tritium decays so fast, then why can we still find the isotope in nature? Tritium is constantly being produced. Nitrogen molecules high up in the atmosphere are bombarded by high-speed cosmic neutrons from cosmic rays. These neutrons make the nitrogen nuclei unstable and they decay to form tritium and carbon.

We have good reason to believe that the Earth is about 4.5 billion years old, but how did we reach this number? The fossil record was a good starting point. Certain fossils can only be found in certain layers, and these layers are always found in the same order. This phenomenon is known as the principle of faunal succession and demonstrates the linear nature of evolution. Fossils can be used to prove that rocks from one part of the world are the same age as rocks in a different part of the world if they contain the same suite of species. However, fossils can only tell us relativistic ages and not absolute ages. In other words, we can say which fossils are older, but not how old they are.

A similar concept is also used in geology. Several geological principles help us to determine how rock layers are formed and the order in which they are laid down. Rock layers can be folded, twisted, buried, eroded, or melted back down and recycled, but the geological principles help to figure out the precise sequence of events, and geologists use observations from the present day to estimate how long it may take for the processes to occur. For example, we can look at the gradual rates of mountain building and erosion, and the movement of the tectonic plates and assume that these rates have not changed significantly since the Earth was young. From this information we can estimate how old the Earth is.

Neither of these methods is particularly good for giving a true age of the Earth. This is where radiometric dating comes in. Carbon dating, or radiocarbon dating, is a highly accurate method for determining how old a rock or fossil is by looking at the proportion of radioactive isotopes that are present compared to non-radioactive isotopes. Isotopes with very long half-lives are preferred because the Earth is very old, and we need to be able to measure rocks that are thousands, millions, and even billions of years old.

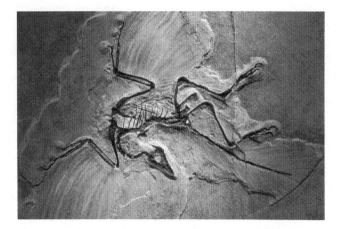

Fig. 21: The "Missing Link." The archeopteryx fossil is estimated to be 150 million years old. Discovered in 1860 in Germany, it is considered by most modern paleontologists to be the missing link between dinosaurs and modern birds.

As the name suggests, carbon is primarily used in carbon dating. Carbon has several isotopes. 12C is the most common and abundant, making up about 98% of all carbon on Earth, followed by 13C making up about 1%. Both of these isotopes have stable nuclei. 14C is a rarer isotope found in only trace amounts but it is radioactive. It has a half-life of approximately 5730 years. This means that every 5730 years, the 14C isotopes will decay by about half. The isotope is constantly being produced in the upper atmosphere as a result of bombardment by neutrons from cosmic rays, and this also allows it to be quite evenly distributed across the whole planet. These three different carbon isotopes make their way into living things through natural processes, and this means that all living things contain trace amounts of 14C which will decay over time.

Carbon dating determines how much of the 14C remains in organic materials to determine how old they are. It can be used for samples between the ages of 500 to about 50,000 years old, after which point there will be too little 14C remaining to give an accurate age. After about 50,000 years, the proportion of 14C will be close to zero.

If we want to determine the age of some seeds found in an archaeological site, we can use 14C. When the seeds formed they absorbed carbon from the environment to build their cells. Over time, the radioactive isotopes of carbon decayed at a known rate. Scientists will compare the amount of 14C to that of a present-day seed and the difference can be used to determine the age. If they only contain half as much 14C as present-day seeds then it can be safely assumed that they are about 5730 years old. After about 10 000 years, we would expect to see only about 30% of the 14C compared to modern day seeds.

For things that are much older than 50 000 years, like rocks, we use different isotopes such as potassium and argon. Radioactive potassium-40 decays to radioactive argon-40. For every 100 atoms of potassium-40 that decay, approximately 11 will become argon-40. The half-life of potassium-40 is immense, at 1.3 billion years, while the half-life of argon-40 is 1.25 million years. Since potassium-40 has such a long half-life and because it is one of the most abundant elements on earth, it's an excellent candidate to date the Earth – this is the secret of how we know Earth is 4.5 billion years old (Tillman, 2021).

In the next chapters we'll move beyond carbon dating and discuss in detail exactly how and why quantum science has shaped our modern world.

23

WE'VE GOT MOST OF SUPERMAN'S SUPERPOWERS NOW (EXCEPT SUPER DISCO ABILITIES)

"So many of our dreams seem impossible, they seem improbable, and then, when we summon the will, they soon become inevitable."

— *SUPERMAN*

Over decades of comics and films, Superman has shown at least fifteen superhuman abilities, to include but not limited to super strength, super speed, super hearing, invulnerability, time travel, freeze breath, super breath, the ability to shoot solar flares, heat vision, laser vision, telescopic vision, microscopic vision, X-ray vision, the ability to turn coal into diamonds, and he even used super disco abilities once to shatter the timing mechanism on a bomb (this is canon, trust me).

While modern humans can't leap over a tall building in a single bound or use laser vision bouncing off a glass mirror to shave, we can use X-ray "vision" to produce highly detailed bone images. When we send X-rays through human tissue in a controlled environment, they pass through the body but some are absorbed by the organs and tissues in different amounts. An image can be created using an X-ray detector which captures all the X-rays that pass through the body. The amount of radiation absorbed by different organs and tissues depends on the **radiological density**. Radiological density is based on the atomic number of the atoms making up different cells and tissues, and their density. Teeth will show up well on an X-ray because they contain relatively heavy elements like calcium, which absorb more X-rays than softer tissues like muscles or fat (Goodman, 2021).

X-rays can be used to identify bone fractures, tumors, calcification, dental problems like cavities, pneumonia of the lungs, and other irregular growths inside the body.

Strange Facts: *At the start of WWI, the French military didn't yet recognize the value of X-rays in treating wounded soldiers - Nobel Prize winner Marie Curie begged for funding to build her "radiological car" but never got it. Finally, the Union of Women from France donated the capital she needed, and her radiological car played a crucial role in treating the wounded in the Battle of Marne in 1914, a major victory for the Allies in defense of Paris.*

She would go on to train 150 women in using "X-ray machines" to treat wounded soldiers in WWI.

THE CURE FOR CANCER

Quantum technology aims to beat cancer. Nuclear medicine, for example, is used to conduct Search and Destroy missions on cancerous cells. One example is the use of "tracers," radiopharmaceuticals that are inserted, ingested, or inhaled into the body. The tracers will move through the body and release gamma rays, which can be detected using special cameras. Doctors can watch the tracers as they travel through different organs or parts of the body to help them diagnose problems and diseases. F-19 fluorodeoxyglucose is a common radiotracer used to help diagnose cancer. Cancer cells are often more metabolically active than healthy cells and will absorb more of this molecule, which acts much like normal glucose. The cancerous cells will emit more gamma radiation as they absorb the tracer.

Radiopharmaceuticals can also be used to treat tumors and cancers because of the way that different cells in the body interact with these molecules. The cells in the thyroid gland, for example, readily absorb iodine from the body. When a person has thyroid cancer, iodine-131, a radioactive isotope, can be ingested and it will be absorbed by the thyroid. The isotope will release gamma radiation that will help to shrink and destroy the cancerous cells over time. Another radioactive isotope, radium-223-dichloride, can be used to treat bone cancer because it closely resembles calcium molecules used the bones to create new tissues. The radiation emitted by the isotope helps prevent cancer development.

Radiation can also be directed into the body from an external beam to target specific areas of the body to help prevent or treat cancer. A machine called a linear accelerator can accurately direct the flow of electromagnetic radiation such as gamma rays or x-rays that will help to break down cancer cells over time (Krans, 2022).

THE HUNT FOR MAN'S MOST PRECIOUS RESOURCE

Quantum science hasn't just enabled us to see inside the body – we can also see underground. This power is particularly useful for finding Earth's most precious resource, water. Radioactive tracers are used to determine the location, volume, and flow-rates of underground water sources and aquifers. It can be extremely difficult to figure out exactly how much water passes through or is stored in an aquifer. Radioactive isotopes make very useful tracers which can be placed at different parts of a spring or aquifer, and then detected at discharge sites or along with other parts of the aquifer. The time it takes for an isotope to move through an aquifer or water body helps scientists estimate the velocity of the water, or flow rate, and from there they can develop more accurate models to determine how much water is located underground. Carbon-14, tritium, and Chlorine-36 are often used for these purposes as they occur naturally and will not contaminate the water sources (Cartwright, 2017).

HOW TO SAVE AN EXTRA 1,100 AMERICANS PER YEAR

According to the National Institute of Standards and Technology, smoke alarms are present in 96% of U.S. homes, but 20% of those homes have non-functioning smoke detectors. They estimate that if smoke detectors were fixed and installed in every household, U.S. fire deaths would drop by about 1,100 *every year*. Smoke detectors use radiation to detect smoke in the air. They contain radioactive elements, usually Americium-241, which emit alpha particles. The alpha particles travel through a chamber where they interact with molecules in the air and strip them of their electrons, creating positively charged ions. The electrons attach themselves to other molecules and create negatively charged ions. The smoke detector is fitted with two electrodes, one with a positive charge and the other with a negative charge. The positively charged ions are attracted to the negatively charged electrode, while the negatively charged ions are attracted to the positively charged electrode, creating a current inside the chamber. When smoke particles enter the chamber they disrupt the current, which triggers an alarm (USNRC, 2017).

ANOTHER INCREDIBLE PREDICTION FROM EINSTEIN IS CONFIRMED TRUE (AGAIN)

In the Superman comics, all inhabitants of the planet Krypton possessed the power of laser vision, but only Superman and

General Zod had completely mastered it. Their cells could absorb solar energy which they could emit through his eyes in the form of a heat laser beam.

As with X-rays, we have innovated ways to create our own modern lasers which have transformed multiple industries, including DVDs, corrective eye surgery, and even self-driving cars.

So how do lasers work? **Lasers**, or Light Amplification by Stimulated Emission of Radiation, are very powerful lights created by harnessing the quantum properties of certain materials such as glass, gas, or crystals. When these materials (called the active medium) are exposed to an electrical current or light source (called the energy source), energy is transferred and causes their electrons to become excited. A typical design uses ruby crystals inside a flash tube with mirrors, called a resonant cavity. When light is emitted into the flash tube, the electrons in the atoms of the ruby crystal become excited by the photons. The electrons can move into higher-energy orbitals around the nucleus if the photons provide energy equal to the difference between the energy levels of the two orbitals. Electrons will eventually return to their ground-state through a process called **spontaneous emission**, during which they will each emit a photon. However, spontaneous emission can take a long time.

Incredibly, in 1916 Einstein predicted that these electrons could be *forced* back to ground state in a process known today as stimulated emission. Einstein proposed that photons prefer to travel together in groups or packets, and that if a material has many atoms that are in an excited state, such as within a ruby struck with light, then it would only take one stray photon of the correct wavelength to stimulate a large number of electrons to emit their own photons as they all return to the ground state together, producing a release of exponentially increasing photons.

Lasers exploit this phenomenon and encourage photon production by using mirrors placed on opposite ends of the tube, one of which is completely reflective and the other of which is only 99% reflective. The photons bounce off the mirrors and excite more of the electrons in the atoms of the ruby crystal, releasing more and more photons. The number of photons bouncing around inside the tube will increase exponentially as long as the ruby crystal is exposed to light or electricity. Finally, a stream of highly focused photons will finally overpower one of the mirrors (which is only 99% reflective) and emerge from the tube as a laser beam (Paschotta, 2008).

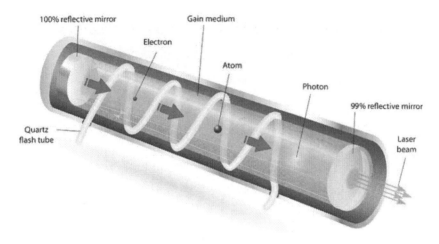

Fig. 22: Laser design.

There are three crucial aspects of lasers that make them so powerful and useful: linewidth, coherence, and power.

1. Lasers have very narrow linewidths, and this means that they contain a very narrow range of frequencies and wavelengths. Most light sources will have a mixture of different frequencies and wavelengths. Sunlight, for example, has many different wavelengths of light ranging from 380 nm to 10 nm. A carbon dioxide will emit light at a wavelength of 10,600 nm, an argon blue laser will emit light with a wavelength of 488 nm, and a Yag laser will emit light at a wavelength of 1,064 nm.
2. Lasers are also coherent, which means that all photons not only have the same wavelength, but they all move in the same direction and are in the same phase. This

means that when visualizing the waves of photons, all of their peaks match up and all of their troughs match up, in the same time and space.
3. Finally, lasers are very powerful and are able to transfer immense amounts of electromagnetic energy onto a very small target. This is what allows them to cut through materials or stay focused on a sharp beam over great distances (Kakalios, 2010).

Lasers are made using crystals such as rubies or other solid materials such as neodymium:yttrium-aluminum-garnet (Yag), which emits infrared light. Gases such as helium or helium-neon mixtures are used to create visible red-light lasers, and carbon dioxide can emit far-infrared light powerful enough to cut hard materials like rocks and metal. Excimer lasers also use gasses, such as chlorine or fluorine, which produce molecules called dimers when stimulated with electricity. The dimers emit ultraviolet light when exposed to electricity (Weschler, 2000).

Strange Facts: *The Race for the world's first laser began in 1958, when two scientists published a paper describing a way to bring Einstein's laser ideas to life with potassium and infrared light. Several top-tier scientist teams around the world, including the Soviet Union, received millions in funding to produce the world's first laser. A young engineering genius named Theodore Maiman, armed with only one part-time assistant and eight months behind the competition, won the race in 1959 with a tiny ruby laser – and he did it with a budget of only $50,000.*

YOU SCRATCHED MY CD

The Britney Spears CD hiding in your closet was made with a laser. CDs, DVDs, and Blu-ray disks are all forms of optical storage, and lasers are used to write and read the information on them. The information is stored using microscopic dots with unique spacing and variations of brightness and darkness, which can be detected and translated using lasers. The laser shines onto these dots and is reflected towards a detector. The laser will read this information using algorithms that convert the spacing, patterns, angles, and other features of the dots into digital data like a song or film (Alleman, 2012).

CARS THAT DRIVE THEMSELVES?

Just like bats can use echolocation to determine their surroundings, self-driving cars use LIDAR to detect surroundings and threats and keep you safe. LIDAR, or Light Detection and Ranging, uses light in a similar way to radar, by sending out short bursts of laser beams and detecting those that return after they reflect and bounce off of objects. LIDAR technology is used to map and measure things in three dimensions and works by measuring the amount of time it takes for a laser beam to reflect onto a receiver device.

Fig. 23: Google's fleet of self-driving Toyota Priuses has now logged more than 190,000 miles driving in traffic and mountainous roads with extremely limited human intervention.

There are so many applications for LIDAR and the equipment can be fitted to stationary detectors, or onto airplanes, drones, small vehicles, and even rovers on Mars. Biologists can use LIDAR devices on airplanes to create highly accurate maps of forests or grasslands, with so much detail that individual leaves on individual trees can be detected. By altering the wavelength of the lasers, LIDAR can penetrate through thick forest canopies to reveal the structure of the forest floor, and this has been used by archaeologists to reveal the hidden networks of ancient cities in places like the Amazon. It is used by autonomous cars to detect obstacles like curbsides, pedestrians, and other barriers which help make self-driving cars a reality. LIDAR is also used in meteorology to predict the weather by measuring clouds, and winds, and it can detect different kinds of emissions from factories, cities, and industries. LIDAR is so useful and adaptable because of the nature of lasers; the wavelength can be altered to suit any needs one may have in measuring and detecting things on the ground, under the oceans, or in the atmosphere (Qi Chen, 2017).

WHERE SHOPPING AND LASERS COLLIDE

Almost every time you purchase something it will have a barcode on the packaging. This is a way to give each item a unique number so that it can be accounted for and so a business can keep track of incoming and outgoing stock. Barcodes are read using barcode scanners, which use lasers. The lasers are emitted from the scanner and the black and white lines on the barcode bounce the light back into a receiver. The white parts of a barcode reflect most light while the black parts reflect very little. The receiver contains photoelectric cells that can detect the incoming light, and a computer can interpret the data. Some barcode scanners use mirrors or rotating prisms, which oscillate the laser beam back and forth to read all parts of the barcode. This works much like binary code, where white patches may translate into 0 while black patches translate into 1, though barcodes generally contain a string of numbers (Woodford, 2018).

WAS BLIND AND NOW I SEE

Much like powerful lasers can be used to cut through metal or rock, they can also be used to replace the surgical scalpel and carry out delicate medical procedures. The high levels of energy that are carried in laser beams can cut through tissues very cleanly and help to minimize damage to the surrounding areas. Lasers are commonly used in corrective eye surgery, removing difficult-to-reach tumors, sealing blood vessels and preventing bleeding, treating growths on the skin, scar removal, tattoo removal, and microsurgery.

An example is Lasik, a procedure in which lasers are used to shave off tissues from the cornea of the eyes to reshape them, helping to improve vision. In Lasik, a laser with a 1mm diameter and 193 nm wavelength is used and blasts the tissues with the energy of 1 microJoule (Shapiro Laser, 2015). Lasers are also used in hair removal by creating a beam with a wavelength that will be absorbed by the pigments inside the hairs. The pigment converts the light into heat which ultimately damages the hair follicles and prevents further hairs from growing. Similarly, tattoos can be removed using lasers. Tattoo pigments, much like the pigments in hair, will absorb certain wavelengths of light and convert the energy into heat. During a tattoo removal session, the laser will be set to those particular wavelengths and the pigment molecules will be broken down by the heat into smaller molecules. Multiple sessions are usually required because the many different colors in a tattoo will need to be targeted by different wavelengths. The wavelengths during laser tattoo removal can range anywhere from 1064 nm to 532 nm. Very quick bursts are used to reduce the potential damage to surrounding tissues. Lasers can also be fitted to flexible fiberoptic devices that can be inserted into the body to access deep tissues and organs for procedures like tumor removal (Janik, 2007).

Various kinds of lasers are used in laser surgeries depending on the type of treatment that is needed. Carbon dioxide lasers are used to treat skin problems as they can easily remove the surface layers without harming deeper tissues (Sardana, 2015). Yag lasers can penetrate deeper tissues and are often used along with flexible fiberoptic devices so that they can be applied directly to internal organs. Lasers can also be used in thermotherapy, to heat certain tissues or organs within the body and this is often used to shrink or destroy tumor cells (Thomas, 2018).

SHINE A LASER ON SOMETHING AND FIGURE OUT WHAT IT IS

Spectroscopy is a wide field that investigates and measures the spectra produced when objects or materials interact with or emit electromagnetic radiation. In laser spectroscopy, lasers are shone onto different materials, and the electromagnetic radiation produced is analyzed to learn more about the composition of the material. Most of the light from a laser will shine onto an object and be reflected at the same wavelength. However, some of the light will also interact with the atoms and molecules, and the photons cause electrons to become excited and jump to higher energy levels. When excited electrons return to their ground state, they emit light at different wavelengths than the laser's wavelength, and this is known as fluorescence. Laser spectroscopy is used to identify the chemical composition of materials because the energy states for different atoms and molecules are unique, and the resulting wavelengths of the emitted light can be correlated to these energy levels. It can be used in agriculture to identify the types of pesticides found on crops by analyzing the resulting fluorescence when a nitrogen laser is shone onto them.

THE REAL-LIFE DEATH STAR

Though the death rays of science fiction remain theoretical, there are many kinds of laser weapons already in use. Most laser weapons are designed to temporarily incapacitate, usually by causing blindness, and are limited in their ability to destroy objects. One of the major limitations of laser weapons is that lasers are only able to cut through hard materials like steel or rock given enough time and power, and short bursts of light do not suffice. In a military setting, many targets that you would want to destroy with a laser beam are moving ones, such as missiles, tanks, aircraft, etc. making it difficult to shine a laser onto them for very long. Weapons developers are working on sophisticated tracking software that will allow lasers to direct their beams onto a target for long periods, allowing the energy to cut through the materials (Zohuri, 2016).

The most pressing challenge for laser weapons, however, is called thermal blooming and is caused by interference from particles in the atmosphere. A laser will have to pass through several miles of the atmosphere before it can hit its target, but even the most precise lasers will experience some energy loss caused by air molecules that absorb the radiation (Atherton, 2021). The surrounding air will heat up and cause further distortions to the laser beam. This is made even worse when there are conditions like rain, snow, fog, smog, or smoke in the atmosphere. Laser weapons would work best under vacuum conditions.

So next time you drive with family and friends to enjoy the next *Star Wars* movie, rest assured the planet-destroying Death Star remains beyond the reach of modern quantum science. The atomic clocks enabling the GPS you may use to navigate to the theater, however, are another matter entirely.

24

THE MOST PRECISE CLOCKS IN THE UNIVERSE

"Lost time is never found again."

— BENJAMIN FRANKLIN

To understand how atomic clocks work, we must first understand what timekeeping is. Astronomical events make for a useful starting point: the Earth rotates around its axis every 24 hours and it completes an orbit around the Sun every 365.25 days. The decision to split a day up into 24 hours is arbitrary and comes from the Ancient Babylonians, who also decided to divide hours and minutes into units of 60. A day begins at sunrise and ends at sunset, which is, on average, about 12 hours. A day could just as easily have been defined in units of 10, or 5, or 7, or 300. However, the tradition persisted and today we define time using this system, where one day is 24 hours, each hour is 60 minutes long, and each minute is 60 seconds long. This means that a second was defined as approximately 1/86400 of a day. This measurement is so small that ancient people didn't really have any use for it and certainly didn't have any way to measure it (Lomb, 2011).

However, the Earth's rotation around its axis is variable and changes very slightly based on tidal interactions, resonant stabilization, and seismic activity. This means that the length of a second based on 1/86400 of a day is also variable. Using these astronomical events, therefore, makes it impossible to keep time accurately.

The first pendulum clocks offered a better way of keeping track of time where the swing of a pendulum serves as a frequency reference that the length of a second is based on. Clock manufacturers had to calibrate their pendulums to all be as similar and accurate as possible to match astronomical time. A more accurate frequency reference comes from quartz crystals, which can resonate at a frequency of 32.768 kHz to 200 MHz, depending on their size and thickness. This is because quartz is piezoelectric, meaning it can convert electrical energy into mechanical vibrations (McCarthy & Seidelmann, 2009). In a quartz clock, electricity is applied to a small quartz crystal which will vibrate. These vibrations are converted into mechanical movements by the gears. The gears use several ratio conversions to slow down the vibrations so that the clock hands can move once per second. Unfortunately, quartz crystals can lose accuracy over time due to temperature or pressure changes, and even when the conditions are stable a quartz clock can gain or lose about 15 seconds every 30 days (Lombardi, 2008).

ENTER THE ATOMIC CLOCK

Modern technology requires much more accurate timekeeping, which is where atomic clocks come into play. **Atomic clocks** work because of the way that energy levels in atoms are quantized. When an atom is exposed to just the right frequency of electromagnetic radiation, it will absorb the energy and its electrons will jump to a higher energy level. Atomic clocks use cesium-133, but all elements have electrons that can make this jump, and each requires energy at very specific frequencies. Rubidium and hydrogen are also used in atomic clocks, and they have resonance frequencies of about 6.8 GHz and 1.4 GHz respectively. This frequency never changes, and in the case of cesium-133, electromagnetic radiation with a frequency of 9,192,631,770 Hz is required. If the frequency is slightly lower, or slightly higher, the quantized nature of the energy levels means that the electrons will not move into a higher orbital.

The resonant frequency of cesium-133 was chosen by the scientific community to represent exactly one second. What this means is that a beam of electromagnetic radiation oscillating at exactly 9,192,631,770 Hz will cause cesium-133 electrons to become excited and move into a higher energy orbital, or put another way, one second is equal to the amount of time it takes for cesium-133 to oscillate exactly 9,192,631,770 times (Samuelson, 2019).

GRAVITY WARPS TIME, SO GPS CLOCKS MUST BE FIXED OFTEN

Atomic clocks are incredibly accurate, especially compared to astronomical time. Many modern systems, such as GPS (global positioning system) rely on this high degree of accuracy to work properly. A GPS device works by measuring the amount of time it takes for a signal to travel to three or more satellites in orbit around the earth. It uses these time measurements to triangulate the exact location that the signals were sent from. GPS satellites need to be able to keep time accurately, as an error of only 1 microsecond can result in you missing your destination by about 300 meters. This is why they use atomic clocks (McCarthy & Seidelmann, 2009).

Another reason that the clocks must be so accurate is because of time dilation. Einstein predicted that if two clocks were placed under different gravitational pulls, that they would tick at different speeds. He called this **time dilation** in his Theory of General Relativity. GPS systems need to overcome this problem, because clocks in orbit run faster by 45,509 ns per day. Over time, these variations will begin to add up and cause problems with navigation. The clocks onboard GPS satellites must therefore be updated twice per day to keep them in sync with the clocks on the surface of the Earth, specifically clocks located at sea level.

Clocks are used in a similar way to keep track of spacecraft and to help them navigate. Signals are sent back and forth between satellites on Earth and in space and the time taken for these signals to travel is measured and used in calculations to determine the heading and distance that has been traveled.

Atomic clocks can be used to test out time dilation on extremely small scales because they are so accurate. In experiments where two atomic clocks were placed a mere 0.04 inches apart, they were still able to detect differences in the Earth's gravitational pull and produced different time readings after about 90 hours. These types of experiments can be used to learn more about the nature of spacetime and how it interacts with gravity (Bothwell, 2022).

THE WORLD'S GREATEST CLOCK

The Deep Space Atomic Clock (DSAC) is one of the most accurate and stable atomic clocks ever developed. DSAC was launched by the SpaceX Falcon Heavy in June 2019 and was activated in August of the same year. It loses no more than 1 ns every 10 days (NASA, 2014).

DSAC is more stable than cesium-133 atomic clocks because it uses positively charged mercury-199 ions. The resonance frequency of these ions is about 40.507 GHz. In a cesium-133 clock the atoms are neutral and contained within a vacuum chamber. There is potential for the atoms to interact with the walls of the vacuum chamber and this can lead to inaccuracies.

However, the positively charged mercury-ions can be held in place using an electromagnetic field, preventing them from interacting with the vacuum chamber, making them much more stable (Samuelson, 2019).

Today, only four world regions retain the use of GPS satellites (and the atomic clocks that enable them): the European Union (EU), Russia, China, and the United States. As such, these regions are capable of sustained warfighting in the sixth military operational domain of modern warfare – space. The fifth operational domain, cyberspace, is the subject of the next chapter.

25

HACKERS FEAR THIS ONE TECHNOLOGY

"If you spend more on coffee than on IT security, you will be hacked. What's more, you deserve to be hacked."

— RICHARD CLARKE

If knowledge is power, data security is its keeper. In the modern world, there is an infinite river of sensitive information ranging from social security numbers, banking details, and internet search history, to industry trade secrets, important financial documents, and top-secret military reports. All of this information is stored as data on computers and servers all over the world. This data needs to be kept safe and secure; it must be protected from hackers and unauthorized parties that want to steal it and use the information for their own purposes. This is why data encryption has become a crucial element of cyber security. Encryption is a method of safeguarding data by encoding it so that it can only be read by someone with a special key. Data can be encrypted using a complex algorithm that scrambles the data before sending it. Once received, the data can only be unscrambled using a matching algorithm, or key.

Strange Facts: *History's first hacker was John Draper, also known as Captain Crunch. In the early 1970s, public telephones were run by an automated system that used specific analogue frequencies to place calls. John was able to make free long distance and international calls by imitating the correct frequencies with free whistles he found in Cap'n Crunch cereal boxes.*

UN-HACKABLE ENCRYPTION?

Traditional encryption methods rely on the special properties of prime numbers. A prime number is an integer that is only divisible by itself and 1, such as 2, 3, 5, 7, 11, 73, 739, 7393, 73939, etc. There are an infinite number of prime numbers, but the largest known is $2^{82,589,933} - 1$, which would have 24,862,048 digits when written out in full. All numbers can be factored into prime numbers, which means that you can figure out which prime numbers can be multiplied together to give you your answer. For example, the prime factorization of 7293 is 3 x 7 x 13 x 17. Prime factorization is extremely difficult and requires a lot of computing power, especially when the numbers become very large.

Traditional encryption relies on this difficulty to secure data. Two large numbers, which are usually prime numbers, are multiplied together and the answer is used to encode a message. To decode the message, we need to know which two prime numbers were multiplied together and there is no easy way to figure this out. However, as computer processing power has become faster and faster, even highly sophisticated methods of encrypting data can be solved when given enough time. To combat this problem, a new type of method is required. Enter **quantum cryptography.** Unlike traditional encryption methods, quantum cryptography is truly un-hackable.

QUANTUM CRYPTOGRAPHY IN ACTION

Let's walk through a quick example of this technology with fictional characters Alice, Bob, and Eve. Let's pretend Alice and Bob are working on a top-secret project that cannot be hacked or intercepted by the conniving Eve. Alice wishes to send Bob a secure message that has been encrypted. She must first send Bob the decryption key that he can use to decrypt the message.

The decryption key (not the message itself) travels from Alice to Bob as a stream of photons which have been polarized through two polarizing filters: a vertical or horizontal filter and a 45-degree left or 45-degree right filter to produce one of only four different possible quantum states.

The polarized photons are then sent down a fiber optic cable to Bob. When Bob receives the stream of photons, he passes them through two beam splitters (horizontal/vertical and 45 degrees left/45 degree right) which "read" and interpret the polarization of each photon. However, Bob has to guess which beam splitter to use, as he does now know how the photons were polarized by Alice. Bob must tell Alice which beam splitter was used for each photon, and then Alice will compare this to which filters she used to polarize each photon. The photons that were read using the wrong beam splitters will be discarded, and everything that is left becomes the decryption key. Bob can then use the decryption key to easily decrypt whatever secret message is sent later.

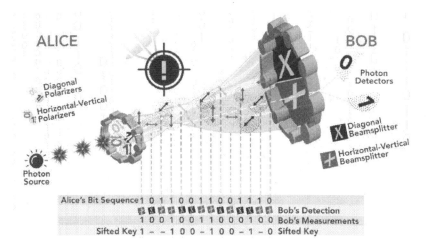

Fig. 23: Quantum Cryptography in action.

If Eve, who wants to steal the sensitive data, can intercept the message, she too can pass the photons through beam splitters. However, there are two problems with trying to intercept the message. First of all, Eve has no idea which filters Alice used to polarize the photons, so there is no way to confirm if her measurements are correct. Secondly, once Eve observes the photons, their quantum states will change, and the change will be detected by both Alice and Bob (Quantum Xchange, 2022).

Today, quantum cryptography is still in its infancy and is yet to be fully developed. There are a few places where it is used and tested, such as Toshiba Corp, the University of Cambridge, The Defense Advanced Research Projects Agency Quantum Network, Quantum Xchange, and some commercial companies. Some of the disadvantages of this method include the fact that photons may be polarized during transit by external factors; the capabilities are limited by the range of fiber optic cables; quantum polarizers and beam splitters are expensive and difficult to use and maintain, and you cannot send out a decryption key to more than one location at a time.

That said, this promising technology has limitless potential – incorruptible voting procedures (and no more voter fraud), unhackable passwords and security systems, and wartime communications that are immune to enemy interception and decryption.

26

THE NEXT STAGE OF COMPUTER EVOLUTION

"Unfortunately, no one can be told what the Matrix is. You have to see it for yourself."

— MORPHEUS, "THE MATRIX"

The digital era of 1s and 0s that exploded on the scene in the 1980s and redefined our modern world will likely remain, but a new quantum era of computer evolution is slowly dawning as traditional computers fail to solve humanity's most complex problems. Even as early as the 1980s, the great physicist Richard Feynman was trying to simulate the electromagnetic interactions between quantum particles but found that classical computers did not have sufficient processing power to run these complex calculations. He realized that the binary system of 1s and 0s was simply not sophisticated enough to complete his work. In a flash of brilliance, he realized that only a quantum computer would be powerful enough for his quantum calculations. He needed a computer that could solve complex problems by operating in multiple states at once, just like the particles he was studying existed in multiple quantum states at once. Just a few years later, the first working quantum computer was developed in 1998 by Isaac Chuang, Neil Gershenfeld, and Mark Kubinec (Lee, 2021).

BITS VS QUANTUM BITS (OR QUBITS)

While traditional computers use binary bits of 1s and 0s, quantum computers use quantum bits, or **qubits**. Classical computers rely on silicon-based chips to store bits, while quantum computers use real subatomic particles such as electrons and photons. A qubit can represent either 1 or 0, but it can also represent a supposition state of both 1 and 0 at the same time, where the probability of observing a 1 or 0 is dependent upon the wave functions of the qubits.

Quantum computers use not only superposition but also entanglement. Quantum entanglement is when two subatomic particles interact with each other in such a way that the state of one will impact the state of the other. For example, two electrons that are entangled always have opposite spin directions, or two photons that are entangled will always have opposite orientations.

Strange Facts: *Einstein described this phenomenon as "spooky action at a distance" because even if the two quantum particles were located on different sides of the galaxy, they could still be entangled, and any change to one would also affect the other instantly. The information travel would be immediate and therefore much, much faster than the speed of light, which only travels 299,792,458 m/s.*

Superposition and entanglement are what make qubits such a powerful computing force. In a classical computer, 1-bit can store either 1 or 0, and 2-bits can store either 00, 01, 10, or 11, and they can only store one of these values at a time. However, a single qubit in a quantum computer can store 0, 1, or both 0 and 1 simultaneously, while 2 qubits can store 00, 01, 10, 11, or store 00, 01, 10, and 11 all at the same time. In other words, n bits can store only one value at a time between 0 and 2^n total possible values, while n qubits can store any value at a time between 0 and 2^n total possible values and *every* value between 0 and 2^n total possible values *at the same time* (Giles, 2019). To put that into perspective, IBM's quantum computer, Eagle, only has 127 qubits, and yet the number of traditional bits a normal computer would need in order to match Eagle's storage would outnumber the number of atoms inside more than 7.5 billion living people. *Just a few hundred cubits could therefore store more pieces of information than there are atoms in the whole universe.* This is the secret of why quantum computers are much, much more powerful than any Mac or PC!

Qubits make it possible to carry out exponentially large calculations. Looking at the problem of encryption from the previous chapter, a classical computer would need millions of years to factorize a large number into its prime components, whereas a quantum computer could solve the same problem in only a few minutes. Even the world's fastest supercomputer, the IBM Summit, is nearly 158 million times slower than the Google quantum computer (Vidar, 2021).

Fig. 24: In 1997, IBM supercomputer Deep Blue defeated Chess World Champion Grandmaster Garry Kasparov in a 6-game match. This historic accomplishment, while incredible, is only the beginning – IBM quantum computers will accomplish much more.

There's one major problem, however, with a computer that relies on superposition to operate. Recall Schrödinger's Cat –

once you try to read the qubit, or in other words, once you open the box to see the cat inside, the superposition of multiple quantum states collapses and the qubit loses its incredible computing powers.

SO HOW CAN WE READ A QUBIT WITHOUT COLLAPSING ITS QUANTUM STATES?

The answer to this riddle is to use quantum entanglement to cleverly observe the qubits indirectly. Instead of looking at the qubit that contains the information, we can read a qubit that it is entangled with and then know the state of the original qubit. One of the ways to do this is to blast superconductors that contain the entangled qubits with short bursts of microwave energy, and then read the signals that are reflected. In other words, if you blast a qubit with energy you can change the spin direction from up to down, and the entangled qubit will also experience a change in spin direction which can be detected without causing the original qubit to collapse from superposition.

Quantum computers have four basic hardware components. These components include the 1) quantum data plane, which contains the physical qubits, 2) the control and measurement plane, which interacts with the qubits and can convert the qubit outputs into digital signals, 3) the control processor plane, which determines the algorithmic sequences used by the control and measurement plane, and 4) the host processor, which handles typical computer functions like user interfaces, operating systems, networks, and storage arrays (National Academies of Sciences, Engineering, and Medicine, 2019).

Inside the quantum data plane, qubits can be encoded with information using logic gates, which assign a specific state to a qubit using electricity or magnetism. An example is a Hadamard Gate, which changes a qubit with a known state to a qubit in a state of superposition. Another type of gate is the Pauli-X Gate, which flips a qubit to an opposite state: a qubit with a 0 value will switch to a 1, and a qubit with a 1 value will switch to a 0 (Anand, 2019).

THE ONE REASON QUANTUM COMPUTERS AREN'T WIDELY USED (YET)

Quantum states, such as superposition, are very fragile, and even tiny changes in temperature, pressure, magnetism, or vibrations, can easily cause qubits to become decoherent. A quantum computer must be able to protect the qubits from any interference or 'noise'. This is why quantum computers must be kept in very strict, controlled conditions within supercooled freezers and vacuum chambers. The temperature must be held at absolute zero (negative 270 °C) where atoms can barely vibrate and only the quantum particles can move (Bonsor & Strickland, 2000). It takes a lot of power to maintain these conditions, which is one of the reasons that quantum computers are not widely used yet.

Along with extremely cold temperatures, there are some other ways that quantum computers can control qubits and protect them from becoming decoherent. One of the common ways to do this is to use trapped-ion qubits, where electrically charged ions are trapped using an electromagnetic field, and a valence electron that can move from a ground state to an excited state acts as the qubit. Qubits can also be controlled using optical traps where light waves control them. Semiconductors with quantum dots can contain and manipulate electron qubits without interfering with them, and superconductors can trap electrons with almost no resistance due to their super-low temperatures (Dideriksen, 2021).

The IBM quantum processors that are used in quantum computers use super-cooled superfluids that can act as superconductors. A **superconductor** is a material that allows electrons to travel through it with no resistance. Cooper pairs, electron pairs that are loosely bound together, can form inside these superconductors. Cooper pairs can jump across barriers like an insulator, using quantum tunneling, and transfer their charge. When two superconductors are placed together but are separated by an insulating material, it forms a Josephson junction. In an IBM quantum processor, the Josephson junctions are used as qubits. Energy can be applied to these qubits to encode them with information or to read their information.

SO WHAT?

Quantum computers will probably not replace classical computers, and you can continue browsing your social media, watching videos, and streaming movies with plain old bits.

That said, since quantum computers are singularly designed for processing large and complex data, they have the potential to change absolutely every industry imaginable. They can model the interactions of subatomic particles. They can be used in biochemistry, where we can create simulations and models of proteins and demonstrate how these long chains of amino acids fold and interact with each other. In medicine, quantum computers may help to predict how molecules behave in the body, accelerating how drugs are developed and potentially making long trial periods unnecessary. They will revolutionize stock market analysis, pattern recognition, machine learning, and perhaps even artificial intelligence networks, which require many complex algorithms to run simultaneously (Das, 2020). Quantum computers are the future.

Regardless of whether or not physicists agree on their interpretations of reality, one thing is certain – we *use* Quantum Physics every day. Quantum technology has built our modern world, and it promises to transform our future. In the final chapter we'll move away from practical science and back, one last time, to the world of theoretical physics to discuss one of the mysterious and exciting topics in all of physics – the multiverse.

27

THE MULTIVERSE AND FINAL THOUGHTS

"You're a man looking at the world through a keyhole. You've spent your life trying to widen it. Your work saved the lives of thousands. What if I told you this reality is one of many?"

— THE ANCIENT ONE, "DR. STRANGE"
(2016)

Parallel universes. Alternate realities. The multiverse. These terms are common knowledge now and are widely used in modern pop culture. In the latest Marvel film, *Dr. Strange and the Multiverse*, a version of Dr. Strange and a young girl named America who has the power to travel through the multiverse are hunted by a demon while searching for the Book of Vishanti. The demon kills this version of Dr. Strange while America escapes through a portal to Earth-616, where she finds an *alternate Dr. Strange* (and you'll have to watch the movie yourself to find out the rest).

But beyond the world of fiction, is the multiverse feasible? Do modern physicists give serious thought to parallel universes or alternate realities?

In short, *they do*. In the third part of this book, we covered Schrödinger's principle of supposition and the thought experiment of a cat in a box. We briefly mentioned the many-worlds interpretation of reality, in which both the live cat and the dead cat exist in separate branches of reality, and when an observer opens the box, he commits to one of the two possible branches and observes either a live cat or a dead cat. But in a separate branch or alternate universe, there is a separate observer who observes the alternate outcome, and there are an infinite number of these branching realities that begin all the time. But they are completely separate and it's impossible to perceive more than one at a time (Stein, 2021).

Strange Facts: *If we ever crossed over into one of these branching realities or alternate universes, what is the probability we would find Life? In 1961, astrophysicist Frank Drake estimated there are 10 advanced civilizations in our Milky Way galaxy alone based on his famous equation, which considers numbers of stars that have planets and numbers of planets that can support Life, among other things. Even with conservative numbers, he predicted* *the universe must be teeming with Life! But enter the Fermi Paradox – if Life is so common, why haven't we found it yet? One answer is that aliens do exist, but either they're not advanced enough to travel to our planet, or they've chosen not to make the trip. One more exciting and frightening possibility is they're en route to our planet but haven't arrived yet. Or perhaps they're already here, but our species denies their existence. As documented footage and sightings of UFOs continue to emerge over the decades from credible sources, however, public opinion on this matter is beginning to shift.*

Critics of the many-worlds interpretation object that, unfortunately, the theory isn't falsifiable. In the traditional scientific method, a theory survives until it is proven wrong by observable and repeatable evidence, and a new theory rises to replace it and so on and so on. But in this case, it's impossible to prove this theory is wrong, just as it's impossible to prove this theory is correct (with today's methods). We will need more evidence in order to prove or deny this theory.

Another competing multiverse theory that many physicists believe says that if the universe is indeed infinitely big, with hundreds of billions of galaxies with trillions and trillions of stars in a vast cosmos that is unimaginably large, there are only so many configurations that a finite number of matter particles can take before these arrangements begin repeating themselves. Eventually, even complex configurations of particles, including entire solar systems and galaxies, must repeat themselves. Supporters of this theory claim there is a strong probability there must be another you in a faraway corner of space, because the universe is *that* large.

A third theory, the **eternal inflation theory**, draws from the history of our own universe for clues and supporting evidence. At the moment of the Big Bang, 13.7 billion years ago, space itself exploded in all directions in a fraction of a second in a process called cosmic inflation, and this occurred even before the explosion of matter and light and atoms that we envision today as the Big Bang. Several researchers today believe that the spontaneous cosmic inflation that birthed our universe is a normal phenomenon that happens all the time in places unimaginably far away, and that when the inflation finally slows, another bubble universe forms. Since the cosmic inflation of space moves faster than the light traveling through space, it would be impossible for us to ever reach one of these bubble universes and they never intersect (Stein, 2021).

Supporters of eternal inflation theory find it attractive because it's simple, and it may help to answer the question, *why are we here?* Theoretical physicist Alexander Vilenkin of Tufts University in Massachusetts is confident that our own existence amid this sea of bubble universes is a result of a "natural selection" of universes: "This picture of the universe, or multiverse, as it is called, explains the long-standing mystery of why the constants of nature appear to be fine-tuned for the emergence of life," Vilenkin writes. "The reason is that intelligent observers exist only in those rare bubbles in which, by pure chance, the constants happen to be just right for life to evolve. The rest of the multiverse remains barren, but no one is there to complain about that."

And so how is it that we find ourselves here in this vast and beautiful universe, filled with stars and galaxies and matter and light and life forms too wondrous and complex to imagine? Whether we choose to believe we are the creations of an intelligent and loving God or that we are here because in the endless non-random natural selection of infinite graveyard universes, *our* universe carried the correct conditions and laws and chemistry to spawn Life, two things are certain. That is, the universe is breathtakingly beautiful. And, we are alive to appreciate it.

ACKNOWLEDGEMENTS AND REVIEWS

My team, Publishing Services, and I have poured our heart and soul into these pages in order to educate, entertain, and inspire our Readers – if you gained Value from reading this book and would like us to publish future books, please let us know by leaving a 5 star review. We appreciate it!

Scan the QR code below to leave your review.

GLOSSARY

Annihilation: reaction in which a particle and its antiparticle collide and disappear, releasing energy. The most common annihilation on Earth occurs between an electron and its antiparticle, a positron.

Antimatter: substance composed of subatomic particles that have the mass, electric charge, and magnetic moment of the electrons, protons, and neutrons of ordinary matter but for which the electric charge and magnetic moment are opposite in sign.

Atom bomb: an explosive weapon that gets its power from a nuclear fisson reaction in which the atomic nuclei of radioactive elements are blasted with energy, causing them to split apart, releasing large amounts of heat and energy.

Atomic clocks: type of clock that uses certain resonance frequencies of atoms (usually cesium or rubidium) to keep time with extreme accuracy.

Black hole: cosmic body of extremely intense gravity from which even light cannot escape. Black holes usually cannot be observed directly, but they can be "observed" by the effects of their enormous gravitational fields on nearby matter.

Chernobyl: nuclear accident that occurred on 26 April 1986 at the No. 4 reactor in the Chernobyl Nuclear Power Plant, near the city of Pripyat in the north of the Ukrainian SSR in the Soviet Union. It is considered the worst nuclear disaster in history both in cost and casualties.

Conservation of energy principle: states energy is neither created nor destroyed. It may transform from one type to another.

Copenhagen Interpretation: a quantum particle doesn't exist in one state or another, but in all of its possible states at once, first posed by physicist Niels Bohr in 1920.

Critical mass: the smallest amount of fissile material needed to sustain a nuclear chain reaction at a controlled, and not exponentially increasing, rate, often used in nuclear reactors.

Dark energy: mysterious and undetected force that appears to be accelerating the rate of universal expansion. Dark energy is used to explain what might be making the universe expand at an ever-increasing rate despite all calculations of known force and matter demonstrating that the rate of expansion should be decelerating.

Dark matter: hypothetical form of matter assumed to exist based only on the movements of stars, planets, and galaxies. It is called 'dark' because it has never been observed and does not interact with electromagnetics, meaning it does not emit, absorb, or reflect electromagnetic radiation such as light - it is invisible. Based on calculations, it appears to account for over 80% of all matter in the universe, because of the gravitational effects that are detected. The leading explanatory theory is WIMPs (weakly interacting massive particles) which may have more than a hundred times the mass of protons but exceptionally weak interactions.

Eternal inflation theory: hypothetical inflationary universe model, which is itself an outgrowth or extension of the Big Bang theory. According to eternal inflation, the inflationary phase of the universe's expansion lasts forever throughout most of the universe.

Event horizon: the boundary surrounding a black hole past which no radiation or matter can escape, and the appearance of the 'black hole' actually begins even though its gravitation effects spread out much further.

General relativity: theory of gravitation developed by Albert Einstein between 1907 and 1915. The theory of general relativity says that the observed gravitational effect between masses results from their warping of spacetime.

Gravitational lensing: as the light emitted by distant galaxies passes by massive objects in the universe, the gravitational pull from these objects can distort or bend the light.

Graviton: hypothetical quantum of gravity, an elementary particle that mediates the force of gravitational interaction.

Ground state: a quantum mechanical state where the energy of an atomic nucleus or atom is lowest. Electrons can move into different energy levels, but the lowest one is the ground state.

Half-life: the time needed for half of the atomic nuclei in a radioactive sample to decay.

Heisenberg's Uncertainty Principle: the position and the velocity of an object cannot both be measured exactly, at the same time, even in theory

Laser (Light Amplification by Stimulated Emission of Radiation): created when electrons in the atoms in optical materials like glass, crystal, or gas absorb the energy from an electrical current or a light. That extra energy "excites" the electrons enough to move from a lower-energy orbit to a higher-energy orbit around the atom's nucleus.

Law of Universal Gravitation: two bodies in space pull on each other with a force directly proportional to their masses and inversely proportional to the square of the distance between them.

Linear superposition: (as opposed to quantum superposition) when two or more wave functions are present at the same time, the disturbance they create is equal to the sum of their parts. Two wave peaks will combine to produce a larger wave peak, which is equivalent to the sum of the original two.

Loop Quantum Gravity: hypothetical quantum theory of gravity based directly on Einstein's geometric formulation rather than the treatment of gravity as a force. As a theory LQG postulates that the structure of space and time is composed of finite loops woven into an extremely fine fabric or network.

Manhattan Project: research and development undertaking during World War II that produced the first nuclear weapons. It was led by the United States with the support of the United Kingdom and Canada

Many-Worlds Interpretation: interpretation of quantum mechanics that asserts that the universal wavefunction is objectively real, and that there is no wave function collapse. This implies that all possible outcomes of quantum measurements are physically realized in some "world" or universe.

M-Theory: a theory in physics that unifies all consistent versions of superstring theory.

Multiverse: hypothetical group of multiple universes. Together, these universes comprise everything that exists: the entirety of space, time, matter, energy.

Nuclear fission: reaction where the nucleus of an atom splits into two or more smaller nuclei, while releasing energy.

Particle-wave duality: concept in quantum mechanics that every particle or quantum entity may be described as either a particle or a wave.

Pauli exclusion principle: states that in a single atom no two electrons will have an identical set or the same quantum numbers (n, l, ml, and ms). To put it in simple terms, every electron should have or be in its own unique state (single state).

Photoelectric effect: a phenomenon in which electrically charged particles are released from or within a material when it absorbs electromagnetic radiation.

Photon: a tiny particle that comprises waves of electromagnetic radiation.

Quanta: the minimum discrete amount of any physical entity involved in an interaction.

Quantum entanglement: a quantum mechanical phenomenon in which the quantum states of two or more objects have to be described with reference to each other, even though the individual objects may be spatially separated.

Quantum cryptography: method of encryption that uses the naturally occurring properties of quantum mechanics to secure and transmit data in a way that cannot be hacked.

Quantum Field Theory (QFT): a theoretical framework that combines classical field theory, special relativity and quantum mechanics.

Quantum Physics: fundamental theory in physics that provides a description of the physical properties of nature at the scale of atoms and subatomic particles.

Quantum tunneling: a phenomenon in quantum mechanics where particles can escape the strong nuclear forces that hold atomic nuclei together, by moving as a wave function. A wave, unlike a particle, can modulate and there is a finite probability that it may be able to cross an energy barrier, allowing a particle to escape the nucleus.

Qubits: a quantum bit. The quantum counterpart of bits of classical computing that serve as the basic unit of information in a quantum computer. Qubits can include the spin direction of an electron, the orientation of a photon, or the electrical charge of ions and atoms.

Radioactive decay: a process where unstable atomic nuclei lose energy in the form of radiation. Materials that have unstable nuclei are radioactive and undergo radioactive decay. Alpha decay, beta decay, and gamma decay are the most common types of radioactive decay where one or more particles, along with energy, are emitted from an unstable atomic nucleus. The nucleus is often transformed or broken down into two or more smaller nuclei in the process.

Radiological density: a measure of how much electromagnetic radiation, usually x-rays, can pass through different parts of the body to create an x-ray image. Areas of high radiological density, such as teeth or bone, will absorb more radiation than areas of low radiological density, such as fat.

Schrödinger's Cat: thought experiment that illustrates a paradox of quantum superposition. In the thought experiment, a hypothetical cat may be considered simultaneously both alive and dead as a result of its fate being linked to a random subatomic event that may or may not occur.

Schrödinger Equation: a linear partial differential equation that governs the wave function of a quantum-mechanical system.

Second Law of Motion: the force F acting on a body is equal to the mass m of the body multiplied by the acceleration a of its center of mass.

Singularity: A gravitational singularity, spacetime singularity or simply singularity is a condition in which gravity is so intense that spacetime itself breaks down catastrophically, such as at the center of a black hole. It is also a point at which some property is infinite. For example, at the center of a black hole, according to classical theory, the density is infinite (because a finite mass is compressed to a zero volume).

Special relativity: explanation of how speed affects mass, time and space. The theory includes a way for the speed of light to define the relationship between energy and matter — small amounts of mass (m) can be interchangeable with enormous amounts of energy (E), as defined by the classic equation $E=mc^2$. Material objects can approach the speed of light but never reach it.

Spontaneous emission: the process where an atom in an excited state spontaneously decays to a lower energy state releasing energy in the form of a photon.

Standard Model of Particle Physics: scientists' current best theory to describe the most basic building blocks of the universe. It explains how particles called quarks (which make up protons and neutrons) and leptons (which include electrons) make up all known matter.

Stimulated emission: the process where an atom in an excited state is triggered to decay to a lower energy state by a photon, which causes the atom to release a second photon with the same frequency and direction as the first photon.

String theory: a theoretical framework in which the point-like particles of particle physics are replaced by one-dimensional objects called strings.

Subcritical mass: an amount of fissile material used in nuclear reactors or nuclear bombs that is not enough to sustain a nuclear chain reaction. A subcritical mass undergoing fission reactions would eventually cease and run out of fuel.

Superconductor: material that achieves superconductivity, which is a state of matter that has no electrical resistance and does not allow magnetic fields to penetrate. An electric current in a superconductor can persist indefinitely. Superconductivity can only typically be achieved at very cold temperatures.

Superposition Principle: the ability of a quantum system to be in multiple states at the same time until it is measured.

Supersymmetry: an extension of the Standard Model that aims to fill some of the gaps. It predicts a partner particle for each particle in the Standard Model.

Theory of Everything: hypothetical, singular, all-encompassing, coherent theoretical framework of physics that fully explains and links together all physical aspects of the universe.

Time dilation: difference in the elapsed time as measured by two clocks. It is either due to a relative velocity between them or to a difference in gravitational potential between their locations. Time slows down relative to outside observers both 1) near objects of great mass and at 2) speeds approaching that of light.

Ultraviolet Catastrophe: prediction of late 19th century/early 20th century classical physics that an ideal black body at thermal equilibrium would emit an infinite quantity of energy, as wavelength decreased into the ultraviolet range. This false prediction was based on the assumption that electromagnetic radiation was continuous. Planck showed instead that electromagnetic radiation can only be release in discrete packets of energy called quanta.

Wormhole (Einstein-Rosen Bridge): speculative structure linking disparate points in spacetime, and is based on a special solution of the Einstein field equations. A wormhole can be visualized as a tunnel with two ends at separate points in spacetime.

EQUATIONS

Eq. 1: Planck's Equation

$$E = nh\nu$$

In this equation,

E = Energy of the electromagnetic wave
n = an integer or a quantum number
h = Planck's Constant
ν = frequency of the light emitted

Eq. 2: Planck's Radiation Law

$$B(v, T) = \frac{2hv^3}{c^2} \frac{1}{\exp(\frac{hv}{k_B T}) - 1}$$

In this equation,

B = spectral radiance of a blackbody
v = frequency
k_B = Boltzmann Constant, 1.380649×10^{-23} m² kg/s²k¹
h = Planck's Constant
c = speed of light
T = absolute temperature

Eq. 3: Einstein's Photon Energy

$$E = KE_{max} + W$$

In this equation,

E = photon energy
KE_{max} = maximum kinetic energy of the freed electron
W = work function

Eq. 4: Photoelectric Effect

$$E = h\nu = KE_{max} + W$$

In this equation,

E = Energy of the electromagnetic wave
h = Planck's Constant
ν = frequency of the light emitted
KE_{max} = maximum kinetic energy of the freed electron
W = work function

Eq. 5: Compton's Effect

$$\lambda' - \lambda = \frac{h}{m_e c}(1 - \cos\theta)$$

In this equation,

λ = initial wavelength
λ' = wavelength after scattering
h = Planck's Constant
m_e = electron mass
c = speed of light
θ = scattering angle, typically 0º, 60º, and 120º

Eq. 6: Einstein's Mass-Energy Equation

$$E = mc^2$$

In this equation,

E = energy

m = mass

c = speed of light

Eq. 7: De Broglie's Wavelength Equation

$$\lambda = \frac{h}{mv}$$

In this equation,

λ = wavelength

h = Planck's constant

m = mass

v = velocity

Eq. 8: Newton's Second Law

$$F = ma$$

In this equation,

F = force
m = mass
a = acceleration

Eq. 9: Schrödinger's Equation

$$\frac{ih(\frac{\partial}{\partial t})\psi(r,\ t)}{2\pi} = \frac{-(h/2\pi)^2}{2m} \nabla^2 \psi(r,\ t) + V(r,t)\psi(r,\ t)$$

In this equation,

i = imaginary number, $\sqrt{-1}$
h = Planck's Constant
$\Psi(r, t)$ = wave function, defined over space and time
m = mass of the particle
∇^2 = Laplacian operator
$V(r, t)$ = potential energy of the particle over space and time

Eq. 10: Heisenberg Uncertainty Principle

$$\Delta p \Delta x \approx \frac{h}{4\pi}$$

In this equation,

Δ = uncertainty
p = momentum
x = position
h = Planck's Constant

Eq. 11: Dirac Equation

$$(\beta mc^2 + c\sum_{n=1}^{3} \alpha_n p_n)\psi(x, t) = \frac{ih\, \partial \psi(x, t)}{2\pi\, \partial t}$$

In this equation,

Ψ = wave function for the electron with spacetime coordinates x, t
p = momentum
c = speed of light
h = Planck's Constant

Eq. 12: Law of Universal Gravitation

$$F = G\frac{m_1 m_2}{r^2}$$

Eq. 11: Law of Universal Gravitation

In this equation,

F = gravitational force between two objects
G = gravitational constant
r = distance between the objects
m_1, m_2 = masses of separate objects

REFERENCES

Aharonov, Y. Cohen, E, Colombo, F & Tollaksen, J. (2017). *Finally Making Sense of the Double Slit Experiment.* https://www.pnas.org/doi/10.1073/pnas.1704649114

Alleman, G. A. (2012). *How DVD Players Work.* HowStuffWorks. https://electronics.howstuffworks.com/dvd-player.htm

Allison, G. (2012). *What Happened to The Soviet Superpower's Nuclear Arsenal? Clues For the Nuclear Security Summit.* HKS Faculty Research Working Paper Series RWP12-038.

Amaldi, U. (2001). *Nuclear Physics From The Nineteen Thirties To The Present Day.* In C. Bernardini & L. Bernardini (Eds.), Enrico Fermi: His Work and Legacy (pp. 151–176). Bologna: Società Italiana Di Fisica: Springer.

Anand, R. (2019). *Crash Course in Quantum Computing Using Very Colorful Diagrams.* Medium. https://towardsdatascience.com/quantum-computing-with-colorful-diagrams-8f7861cfb6da

Atherton, K. D. (2021). *General Atomics and Boeing Will Build A Giant Laser For The US Military.* Popular Science. https://www.popsci.com/technology/military-defensive-laser-weapon/

Aziz, A (2020). Schrödinger*'s Wave Equation: Derivation & Equation.* https://www.electrical4u.com/schrodinger-wave-equation/

Bamford, T. (2020). *The Most Fearsome Sight: The Atomic Bombing of Hiroshima.* The National WWII Museum | New Orleans. https://www.nationalww2museum.org/war/articles/atomic-bomb-hiroshima

Belendez, A. (2015). *Faraday and the Electromagnetic Theory Of Light.* https://www.bbvaopenmind.com/en/science/leading-figures/faraday-electromagnetic-theory-light/

Bennet, Jay (2016). *The One Theory of Quantum Mechanics that Actually Kind of Makes Sense.* Popular Mechanics. https://www.popularmechanics.com/space/a24114/pilot-wave-quantum-mechanics-theory/

Bernstein, J. (2010). *Nuclear Weapons: What You Need to Know.* Cambridge Univ. Press.

Betz, E (2020). *What's the Difference Between Dark Matter and Dark Energy?* https://astronomy.com/news/2020/03/whats-the-difference-between-dark-matter-and-dark-energy

Blackbody Radiation and Planck's Quantum Hypothesis. CK12. https://www.ck12.org/book/ck-12-physics---intermediate/section/23.1/

Bonolis, L. (2001). Enrico Fermi's Scientific Work. In C. Bernardini & L. Bonolis (Eds.), *Enrico Fermi: His Work and Legacy.* (pp. 314–394). Bologna: Società Italiana Di Fisica: Springer.

Bonsor, K., & Strickland, J. (2000). *How Quantum Computers Work.* HowStuffWorks. https://computer.howstuffworks.com/quantum-computer1.htm

Bothwell, T., Kennedy, C. J., Aeppli, A., Kedar, D., Robinson, J. M., Oelker, E., Staron, A., & Ye, J. (2022). *Resolving the Gravitational Redshift Across A Millimetre-Scale Atomic Sample.* Nature, 602(7897), 420–424. https://doi.org/10.1038/s41586-021-04349-7

Bova, F., Goldfarb, A., & Melko, R. (2021). *Quantum Computing Is Coming. What Can It Do?* Harvard Business Review. https://hbr.org/2021/07/quantum-computing-is-coming-what-can-it-do

Bradben, Sonia. (2022). *Quantum Computing History and Background.* Azure Quantum. https://docs.microsoft.com/en-us/azure/quantum/concepts-overview

Briggs, A., Whittz, K. (2022). *What Are Black Holes?* Earthsky. https://earthsky.org/space/definition-what-are-black-holes/

Briggs, A. (2020). *What Is Dark Matter?* Earthsky. https://earthsky.org/astronomy-essentials/definition-what-is-dark-matter/

Briggs, A. (2020). *What is Dark Energy?* Earthsky. https://earthsky.org/space/definition-what-is-dark-energy/

Brubaker, B (2021). *How Bell's Theorem Proved "Spooky Action at A Distance Is Real."* https://www.quantamagazine.org/how-bells-theorem-proved-spooky-action-at-a-distance-is-real-20210720/

Burton, K (2020). *Destroyer Of Worlds: The Making of An Atomic Bomb.* https://www.nationalww2museum.org/war/articles/making-the-atomic-bomb-trinity-test

Byrd. D. (2015). *First Photo of Light as a Particle and a Wave.* Earthsky.

https://earthsky.org/human-world/first-photo-of-light-as-both-particle-and-wave/

Caltech. *What Is Quantum Physics?* Science Exchange. https://scienceexchange.caltech.edu/topics/quantum-science-explained/quantum-physics

Cartwright, I., Cendón, D., Currell, M., & Meredith, K. (2017). *A Review of Radioactive Isotopes And Other Residence Time Tracers In Understanding Groundwater Recharge: Possibilities, Challenges, And Limitations.* Journal of Hydrology, 555, 797–811. https://doi.org/10.1016/j.jhydrol.2017.10.053

Clavin, W (2020). *Where is Dark Matter Hiding?* Caltech. https://magazine.caltech.edu/post/where-is-dark-matter-hiding

Cofield, C. (2017). *600-Year-Old Starlight Bolsters Einstein's "Spooky Action at a Distance."* Space.com. https://www.space.com/35676-einstein-spooky-action-starlight-quantum-entanglement.html

Coles, P. (2019). *Einstein, Eddington And The 1919 Eclipse.* Nature, 568(7752), 306–307. https://doi.org/10.1038/d41586-019-01172-z

Coolman, R. (2014). *What Is Quantum Mechanics?* Live Science. https://www.livescience.com/33816-quantum-mechanics-explanation.html

Das, N. (2022). *Pauli Exclusion Principle: Definition, Application, Explanation and Explanation.* College Dunia. https://collegedunia.com/exams/pauli-exclusion-principle-definition-application-explanation-examples-chemistry-articleid-532

Das, S. (2020). *Top Applications of Quantum Computing Everyone Should Know About.* Analytics India Magazine. https://analyticsindiamag.com/top-applications-of-quantum-computing-everyone-should-know-about/

Dattaro, L. (2018). *The Quest to Test Quantum Entanglement.* Symmetry Magazine. https://www.symmetrymagazine.org/article/the-quest-to-test-quantum-entanglement

David M. Hanson, Harvey, E, Sweeney, R & Zielinski, T. (2016). *The Schrödinger Wave Equation for the Hydrogen Atom.* https://chem.libretexts.org/Courses/University_of_California_Davis/UCD_Chem_107B%3A_Physical_Chemistry_for_Life_Scientists/Chapters/4%3A_Quantum_Theory/4.10%3A_The_Schr%C3%B6dinger_Wave_Equation_for_the_Hydrogen_Atom

Dideriksen, K. B., Schmieg, R., Zugenmaier, M., & Polzik, E. S. (2021). *Room-*

Temperature Single-Photon Source With Near-Millisecond Built-In Memory. Nature Communications, 12(1). https://doi.org/10.1038/s41467-021-24033-8

Baird, C. (2013). *What Did Schrödinger's Cat Experiment Prove?* WTAMU. https://www.wtamu.edu/~cbaird/sq/2013/07/30/what-did-schrodingers-cat-experiment-prove/

Dusto, A (2020). *De Broglie Wavelength: Definition, Equation and How to Calculate*. Sciencing. https://sciencing.com/de-broglie-wavelength-definition-equation-how-to-calculate-13722583.html

Efthimiades, S. *Physical Meaning and Derivation of Schrödinger and Dirac's Equations*. Arxiv. https://arxiv.org/vc/quant-ph/papers/0607/0607001v1.pdf

Faizi, S & Fisher, C. (2021). *Pauli Exclusion Principle*. https://chem.libretexts.org/Bookshelves/Physical_and_Theoretical_Chemistry_Textbook_Maps/Supplemental_Modules_(Physical_and_Theoretical_Chemistry)/Electronic_Structure_of_Atoms_and_Molecules/Electronic_Configurations/Pauli_Exclusion_Principle

Fore, M. (2020). *Pauli Exclusion Principle: What Is It & Why Is It Important?* Sciencing. https://sciencing.com/spin-quantum-number-definition-how-to-calculate-significance-13722569.html

Francis, Mathew (2011). *What Does the Double Slit Experiment Actually Show?* Scientific American. https://blogs.scientificamerican.com/guest-blog/what-does-the-new-double-slit-experiment-actually-show/

Freudenrich, C & Kiger, P. (2022). *How Nuclear Bombs Work*. How Stuff Works. https://science.howstuffworks.com/nuclear-bomb.htm#pt2

Friedman J, Patel, V., Chen, W. (2000). *Quantum Superposition of Distinct Macroscopic States*. Nature. https://www.nature.com/articles/35017505#citeas

Frohlich, J. (2017). *Why We Can Stop Worrying and Love the Particle Accelerator*. Aeon. https://aeon.co/ideas/why-we-can-stop-worrying-and-love-the-particle-accelerator

Gamow, G. (1928). *Zur Quantentheorie Des Atomkernes*. Zeitschrift Fur Physik, 51(3-4), 204–212. https://doi.org/10.1007/bf01343196

George, A. (2020). *The Four Fundamental Forces of Nature*. Clearias. https://www.clearias.com/four-Fundamental-forces-of-nature/

Gerritsma, R, Kirchmair, G, Zahringer, F, Solano, E, Blatt R, Roos, F. (2009, September 3). *Quantum Simulation of Dirac Equation.* Retrieved from https://arxiv.org/abs/0909.0674

Gerritsma, R, Kirchmair, G, Zahringer, F, Solano, E, Blatt R, Roos, F. (2009, September 3). *Quantum Simulation of Dirac Equation.* Retrieved from https://arxiv.org/abs/0909.0674

Giles, M. (2019). *Explainer: What Is a Quantum Computer?* MIT Technology Review. https://www.technologyreview.com/2019/01/29/66141/what-is-quantum-computing/

Gillies, J. (2011). *Luminosity? Why Don't We Just Say Collision Rate?* Cern. https://home.cern/news/opinion/cern/luminosity-why-dont-we-just-say-collision-rate

Goel, A. & Hoang, T. (2014). Compton Effect. Radiopaedia. https://radiopaedia.org/articles/compton-effect

Goodman, E. (2021). *Types Of Radiation Therapy: How They Work and What To Expect.* Medical News Today. https://www.medicalnewstoday.com/articles/types-of-radiation-therapy

Goodwin, Z. (2022). *What is Schrödinger's Equation and How is It Used?* Physlink.com. https://www.physlink.com/education/askexperts/ae329.cfm

Groves, L. R. (1975). *Now It Can Be Told: The Story Of The Manhattan Project.* Da Capo Press.

Harari, Y. (2011). *Sapiens: A Brief History of Mankind.* Vintage Books.

Harris, W. and Freudenrich, C. (2000). *How Light Works.* How Stuff Works. https://science.howstuffworks.com/light6.htm

Harris, W. (2021). *How Newton's Laws of Motion Work.* How Stuff Works. https://science.howstuffworks.com/innovation/scientific-experiments/newton-law-of-motion3.htm

Higgins, A. (2009). *Atom Smasher Preparing 2010 New Science Restart.* US News. https://www.usnews.com/science/articles/2009/12/18/atom-smasher-preparing-2010-new-science-restart

Hilgevoord, J. (2001). *The Uncertainty Principle.* Stanford. https://plato.stanford.edu/entries/qt-uncertainty/

Editors (2017). *Manhattan Project.* History.com. https://www.history.com/topics/world-war-ii/the-manhattan-project

Hooper, D. (2020). *How Schrödinger Failed to Find Flaws in the Copenhagen Interpretation*. Retrieved from https://www.thegreatcoursesdaily.com/how-einstein-failed-to-find-flaws-in-the-copenhagen-interpretation/

Hossenfelder, Sabine (2020). *David Bohm's Pilot Wave Interpretation of Quantum Mechanics*. Back Reaction. http://backreaction.blogspot.com/2020/10/david-bohms-pilot-wave-interpretation.html

Howell, E. (2017). *Photoelectric Effect: Explanation and Applications*. Live Science. https://www.livescience.com/58816-photoelectric-effect.html

Huygens, C. (1690). *Treatise on Light*. St. Marys. https://www.stmarys-ca.edu/sites/default/files/attachments/files/Treatise_on_Light.pdf

Janik, J. P., Markus, J. L., Al-Dujaili, Z., & Markus, R. F. (2007). *Laser Resurfacing*. Seminars in Plastic Surgery, 21(3), 139–146. https://doi.org/10.1055/s-2007-991182

Jaupi, J. (2022). *How Does a Nuclear Bomb Work?* Sun.com. https://www.the-sun.com/tech/4790011/nuclear-bomb-atomic-hydrogen-explosion/

Johnson, L. (*2020*). *Schrödinger's Equation: Explained and How to Use It*. Sciencing. https://sciencing.com/schrodingers-equation-explained-how-to-use-it-13722578.html

Kakalios, J. (2010). *The Amazing Story Of Quantum Mechanics*. Penguin Group (USA) Inc. https://xn--webducation-dbb.com/wp-content/uploads/2020/05/James-Kakalios-The-Amazing-Story-of-Quantum-Mechanics_-A-Math-Free-Exploration-of-the-Science-that-Made-Our-World-Gotham-2010.pdf

Kelly, C. (2005). *Remembering the Manhattan Project*. World Scientific. https://www.worldscientific.com/worldscibooks/10.1142/5654

Kragh, H. (2020). *Pauli's Exclusion Principle: The Origin and Validation of a Scientific Principle*. University of Chicago. https://www.journals.uchicago.edu/doi/abs/10.1093/bjps/axn056?journalCode=bjps

Krans, B. (2022). *Radiation Therapy: Side Effects, Purpose, Process, And More*. Healthline. https://www.healthline.com/health/radiation-therapy

Lee, M. (2021). *History Of Quantum Computing: Simplified*. Q-munity Tech. https://www.qmunity.tech/post/history-of-quantum-computing-simplified

Lincoln, D. (2014). *What Are Gravitons?* PBS. https://www.pbs.org/wgbh/nova/article/what-are-gravitons/

Ling, S. J., Sanny, J., Moebs, W. (2016). Openstax College, & Rice University.

Lombardi, M. (2008). *The Accuracy and Stability of Quartz Watches*. Horological Journal. https://tf.nist.gov/general/pdf/2276.pdf

Madhu. (2019). *Difference Between Quantum Physics and Particle Physics*. Difference Between. https://www.differencebetween.com/difference-between-quantum-physics-andparticlephysics/#Quantum%20Physics%20vs%20Particle%20Physics%20in%20Tabular%20Form

Mann, A. (2019). *What is Dark Energy?* Live Science. https://www.livescience.com/what-is-dark-energy.html

Mann, A. (2020). *What is Dark Matter?* Live Science. https://www.livescience.com/dark-matter.html

Mann, A. (2020). *Schrödinger's Cat: The Favorite, Misunderstood Pet of Quantum Mechanics*. Live Science. https://www.livescience.com/schrodingers-cat.html

Marrison, W. A. (1948). *The Evolution of the Quartz Crystal Clock*. Bell System Technical Journal, 27(3), 510–588. https://doi.org/10.1002/j.1538-7305.1948.tb01343.x

Matthews, R. *What Are Gravitons, and Do They Really Exist?* Science Focus. https://www.sciencefocus.com/science/what-are-gravitons-and-do-they-really-exist/

McCarthy, D. D., & Seidelmann, P. K. (2009). *Time - From Earth Rotation to Atomic Physics*. Wiley. https://doi.org/10.1002/9783527627943

McKie, R. (2019). *100 Years On: The Pictures That Changed Our View Of The Universe*. The Guardian. https://www.theguardian.com/science/2019/may/12/100-years-on-eclipse-1919-picture-that-changed-universe-arthur-eddington-einstein-theory-gravity

McMillan, E. (2021). Manhattan Project. Britannica. https://www.britannica.com/event/Manhattan-Project

Melkikh, A. (2021). *Quantum System: Wave Function, Entanglement and the Uncertainty Principle*. World Scientific. https://www.worldscientific.com/doi/pdf/10.1142/S0217984921502225

Merali, Z. (2020). *This Twist on Schrödinger's Cat Paradox has Major Implications for Quantum Theory*. Scientific American. https://www.scientificamerican.com/article/this-twist-on-schrodingers-cat-paradox-has-major-implications-for-quantum-theory/

Merzbacher, E. (2002). *The Early History of Quantum Tunneling.* Physics Today, 55(8), 44–49. https://doi.org/10.1063/1.1510281

Michaelsen Lyn, S. (2020). *Paul Dirac: The Most Beautiful Equation.* Suzette Lyn. https://suzettelyn.com/articles/paul-dirac-the-most-beautiful-equation/

M-Theory. Wikipedia. https://simple.wikipedia.org/wiki/M-theory

Nathan, M. (2020). *The History of the Atomic Model: Schrödinger and the Wave Equation.* https://www.breakingatom.com/learn-the-periodic-table/the-history-of-the-atomic-model-schrodinger-and-the-wave-equation

National Academies of Sciences, Engineering, and Medicine. (2019). *5 Essential Hardware Components of a Quantum Computer. In Quantum Computing: Progress and Prospects* (pp. 113–124). National Academies Press. https://nap.nationalacademies.org/read/25196/chapter/7

Nazario, B. (2010). *What To Expect From Radiation Therapy.* WebMD. https://www.webmd.com/cancer/what-to-expect-from-radiation-therapy

Parry-Hill, M. & W. Davidson, M. (2017). *Thomas Young's Double Slit Experiment.* https://micro.magnet.fsu.edu/primer/java/interference/doubleslit/

Paschotta, R. (2008). *Field Guide to Lasers.* Spie Press.

Piper, K. (2019). *The $22 Billion Gamble: Why Some Physicists Aren't Excited About Building a Bigger Particle Collider.* Vox. https://www.vox.com/future-perfect/2019/1/22/18192281/cern-large-hadron-collider-future-circular-collider-physics

Planck Solves the Ultraviolet Catastrophe [PDF file]. Retrieved from https://www.webassign.net/question_assets/buelemphys1/chapter27/section27dash1.pdf

Pratt, Carl J. (2021). *Quantum Physics for Beginners.* Ippoceronte Publishing.

Qi, Chen. (2017). LIDAR *Remote Sensing and Applications.* Taylor & Francis Ltd.

Quantization of Energy. Retrieved from https://chem.libretexts.org/Courses/Howard_University/General_Chemistry%3A_An_Atoms_First_Approach/Unit_1%3A__Atomic_Structure/Chapter_2%3A_Atomic_Structure/Chapter_2.2%3A__Quantization_of_Energy

Quantum Field Theory (2017). Retrieved from http://www.quantumphysicslady.org/glossary/quantum-field-theory/

Quantum Field Theory. (2020). Retrieved from https://plato.stanford.edu/entries/quantum-field-theory/

Quantum Mechanical Atomic Model (2021). Retrieved from https://flexbooks.ck12.org/cbook/ck-12-chemistry-flexbook-2.0/section/5.11/primary/lesson/quantum-mechanical-atomic-model-chem/

Quantum Xchange (2022). Retrieved from https://quantumxc.com/blog/quantum-cryptography-explained/

Rabbitt, S. (2017). *The Manhattan Project*. Atomic Heritage Project. https://www.atomicheritage.org/history/manhattan-project

Rasmussen, J. (2000). *Applications of Radioactivity*. Britannica. https://www.britannica.com/science/radioactivity/Applications-of-radioactivity

Redd, N. T. (2017). *What Is Wormhole Theory?* Space.com. https://www.space.com/20881-wormholes.html

Rhodes, R. (2012). *The Making of The Atomic Bomb*. Simon & Schuster.

Rivest, R. L., Shamir, A., & Adleman, L. (1978). *A Method for Obtaining Digital Signatures And Public-Key Cryptosystems*. Communications of the ACM, 21(2), 120–126. https://doi.org/10.1145/359340.359342

Samuelson, A. (2019). *What Is an Atomic Clock?* (J. Nelson, Ed.). NASA. https://www.nasa.gov/feature/jpl/what-is-an-atomic-clock

Sardana, K., Ranjan, R., & Ghunawat, S. (2015). *Optimizing Laser Tattoo Removal*. Journal of Cutaneous and Aesthetic Surgery, 8(1), 16. https://doi.org/10.4103/0974-2077.155068

Saul, L. (2018). *The Different Types of Lasers and Lasing Media*. Azooptics. https://www.azooptics.com/Article.aspx?ArticleID=1346

Seigel, E (2019). *This Is Why Quantum Field Theory is More Fundamental Than Quantum Mechanics*. Forbes. https://www.forbes.com/sites/startswithabang/2019/04/25/this-is-why-quantum-field-theory-is-more-Fundamental-than-quantum-mechanics/

Sherrill, D. (2006). The Ultraviolet Catastrophe. Retrieved from http://vergil.chemistry.gatech.edu/notes/quantrev/node3.html

Singh, A. & Quach, Tu. (2020). *Wave-Particle Duality*. Retrieved from https://chem.libretexts.org/Bookshelves/Physical_and_Theoretical_Chemistry_Textbook_Maps/Supplemental_Modules_(Physical_and_Theoretical_Chemistry)/Quantum_Mechanics/02.

_Fundamental_Concepts_of_Quantum_Mechanics/Wave-Particle_Duality

Singh, B., Singh, J. & Kaur, A. (2013). *Applications of Radioisotopes in Agriculture*. Stanford. http://large.stanford.edu/courses/2017/ph241/white-m2/docs/singh.pdf

Skuse, B. (2018). *The Basics of Black Holes Explained*. Sky at Night Magazine. https://www.skyatnightmagazine.com/space-science/the-basics-of-black-holes-explained/

Smith, H. (2018). *What Is a Black Hole?* NASA.gov. https://www.nasa.gov/audience/forstudents/k-4/stories/nasa-knows/what-is-a-black-hole-k4.html

Soutter, W. (2012). *How Do Lasers Work?* Azooptics. https://www.azooptics.com/Article.aspx?ArticleID=368

Stanley, M. (2019). *The Man Who Made Einstein World-Famous*. BBC News. https://www.bbc.com/news/science-environment-48369980

Stein, V. (2021). *Do Parallel Universes Exist? We Might Live in A Multiverse*. Space.com. https://www.space.com/32728-parallel-universes.html

Tyson, N. (2020). *String Theory Explained: A Basic Guide to String Theory*. Masterclass. https://www.masterclass.com/articles/string-theory-explained#what-is-string-theory

Sutter, P. (2020). *String Theory vs. M-Theory: A Showdown to Explain our Universe*. Space.com. https://www.space.com/string-theory-11-dimensions-universe.html

Sutter, P. (2021). *What is Quantum Entanglement?* Live Science. https://www.livescience.com/what-is-quantum-entanglement.html

Sutter, P. (2021). *What are wormholes?* Live Science. https://www.livescience.com/what-are-wormholes

Swarthout, B. (2014). *What Was Planck's Mathematical Trick to Formulate the Quantum Hypothesis?* Socratic.org. https://socratic.org/questions/what-was-the-plank-s-mathematical-trick-to-formulate-the-quantum-hypothesis

Taskila, O. (2017). *Why Laser Wavelengths Matter for Removing Tattoos*. Info.astanzalaser.com. https://info.astanzalaser.com/blog/why-laser-wavelengths-matter-for-removing-tattoos

The Compton Effect. Retrieved from https://opentextbc.ca/university physicsv3openstax/chapter/the-compton-effect/

The De Broglie-Bohm Pilot Wave Theory. Retrieved from http://flax.nzdl.org/greenstone3/flax?a=d&c=BAWEPS&d=D1206&dt=simple&p.a=b&p.s=ClassifierBrowse

The Heisenberg's Uncertainty Principle. Retrieved from https://opentextbc.ca/universityphysicsv3openstax/chapter/the-heisenberg-uncertainty-principle/

The Photoelectric Effect. Retrieved from https://www.ux1.eiu.edu/~cfadd/1160/Ch28QM/Photo.html

The Schrödinger Equation in One Dimension. Retrieved from https://faculty.chas.uni.edu/~shand/Mod_Phys_Lecture_Notes/Chap7_Schrodinger_Equation_1D_Notes_s12.pdf

Thomas, L. (2018). *How Lasers Are Used for Precise Surgical Work*. AZoM.com. https://www.azom.com/article.aspx?ArticleID=15914

Thompson, H. & Havern, S. *The History of Gravity*. Stanford. https://web.stanford.edu/~buzzt/gravity.html

Tillman, N & Biggs, B (2021). *What Are Black Holes? Facts, Theory and Definition*. Space.com. https://www.space.com/15421-black-holes-facts-formation-discovery-sdcmp.html

Tillman, N. (2013). *What is Dark Energy?* Space.com. https://www.space.com/20929-dark-energy.html

Tillman, N (2022). *What Is Dark Matter?* Space.com. https://www.space.com/20930-dark-matter.html

Tillman, N, Bartels M. & Dutfield, S. (2022). *Eintein's Theory of General Relativity*. Space.com. https://www.space.com/17661-theory-general-relativity.html

Tretkoff, E. (2008). *Thomas Young and the Nature of Light*. APS.org. https://www.aps.org/publications/apsnews/200805/physicshistory.cfm

Ultraviolet Catastrophe – Rayleigh-Jeans Catastrophe. Retrieved from https://www.nuclear-power.com/nuclear-engineering/heat-transfer/radiation-heat-transfer/ultraviolet-catastrophe-rayleigh-jeans-catastrophe/

United States Nuclear Regulatory Commission. (2008). *NRC: Backgrounder on Chernobyl Nuclear Power Plant Accident*. Nrc.gov. https://www.nrc.gov/reading-rm/doc-collections/fact-sheets/chernobyl-bg.html

Urone, P. & Hinrichs, R. (2012). *Newton's Universal Law of Gravitation.* Open Stax. https://openstax.org/books/college-physics/pages/6-5-newtons-universal-law-of-gravitation

USNRC. (2017). *NRC: Backgrounder On Smoke Detectors.* Nrc.gov.. https://www.nrc.gov/reading-rm/doc-collections/fact-sheets/smoke-detectors.html

Vidar. (2021). *Google's Quantum Computer Is About 158 Million Times Faster Than The World's Fastest Supercomputer.* Medium. https://medium.com/predict/googles-quantum-computer-is-about-158-million-times-faster-than-the-world-s-fastest-supercomputer-36df56747f7f

Walchover, N. (2018). *Famous Experiment Dooms Alternative To Quantum Weirdness.* Quanta Magazine. https://www.quantamagazine.org/famous-experiment-dooms-pilot-wave-alternative-to-quantum-weirdness-20181011/

Waltar, A. (2003). *The Medical, Agricultural and Industrial Benefits of Nuclear Technology.* L.A. Radioactive. https://www.laradioactive.com/site/pages/RadioPDF/Waltar.pdf

Wasser, L. A. (2020). *The Basics of Lidar - Light Detection and Ranging - Remote Sensing.* Neon Science. https://www.neonscience.org/resources/learning-hub/tutorials/lidar-basics

Wave-Particle Duality. Retrieved from https://www2.nau.edu/~gaud/bio302/content/clsphy.htm

Webb, R. (2008). *A Relative Success.* Nature. https://www.nature.com/articles/milespin04

Webb, R. *Quantum Field Theory.* New Scientist. https://www.newscientist.com/definition/quantum-field-theory/

Weiss, U., Biber, P., Laible, S., Bohlmann, K., & Zell, A. (2010). *Plant Species Classification Using A 3D LIDAR Sensor and Machine Learning.* Ninth International Conference on Machine Learning and Applications (ICMLA).

Wells, S (2021). *Dark Energy: The Physics – Breaking Force that Shapes Our Universe, Explained.* Inverse. https://www.inverse.com/science/dark-matter-energy-explained

Weschler, M. (2000). *How Lasers Work.* How Stuff Works. https://science.howstuffworks.com/laser.htm

Wilde, R. (2019). *What Is the Theory Behind Mutually Assured Destruction?* ThoughtCo. https://www.thoughtco.com/mutually-assured-destruction-1221190

Wiseman, H (2014). *Bell's Theorem Still Reverberates.* Nature. https://www.nature.com/articles/510467a

Witman, S (2014). *Ten Things You Might Not Know about Particle Accelerators.* Symmetry. https://www.symmetrymagazine.org/article/april-2014/ten-things-you-might-not-know-about-particle-accelerators

Wood, C. & Stein, V. (2022). *What Is String Theory?* Space.com. https://www.space.com/17594-string-theory.html

Woodford, C. (2018). *How Do Barcodes and Barcode Scanners Work?* Explain That Stuff. https://www.explainthatstuff.com/barcodescanners.html

Woods, S. & Baumgartner, K. (2021). *Heisenberg's Uncertainty Principle.* Retrieved from https://chem.libretexts.org/Bookshelves/Physical_and_Theoretical_Chemistry_Textbook_Maps/Supplemental_Modules_(Physical_and_Theoretical_Chemistry)/Quantum_Mechanics/02._Fundamental_Concepts_of_Quantum_Mechanics/Heisenberg's_Uncertainty_Principle

World Nuclear Association. (2021). *Chernobyl.* World Nuclear Association. https://world-nuclear.org/information-library/safety-and-security/safety-of-plants/chernobyl-accident.aspx

Zimmerman Jones, A. & Robbins, D. (2016). *String Theory and Loop Quantum Gravity.* Dummies. https://www.dummies.com/article/academics-the-arts/science/physics/string-theory-and-loop-quantum-gravity-177738

Zimmerman Jones, A. & Robbins D. (2016). *Einstein's Special Relativity.* Dummies. https://www.dummies.com/article/academics-the-arts/science/physics/einsteins-special-relativity-193336

Image Credits

All images are sourced from Wikimedia Commons.

Made in the USA
Monee, IL
06 January 2023

24646861R00127